METHODS IN MOLECULAR BIOLOGY™

T0180455

Series Editor
John M. Walker
School of Life Sciences
University of Hertfordshire
Hatfield, Hertfordshire, AL10 9AB, UK

For other titles published in this series, go to
www.springer.com/series/7651

Microengineering in Biotechnology

Edited by

Michael P. Hughes and Kai F. Hoettges

University of Surrey, Guildford, UK

Editors
Michael P. Hughes
University of Surrey
Centre for Biomedical
   Engineering
Guildford
Duke of Kent Building
United Kingdom GU2 7TE
m.hughes@surrey.ac.uk

Kai F. Hoettges
University of Surrey
Centre for Biomedical
   Engineering
Guildford
Duke of Kent Building
United Kingdom GU2 7TE
k.hoettges@surrey.ac.uk

ISSN 1064-3745
ISBN 978-1-61779-640-1   e-ISBN 978-1-60327-106-6
DOI 10.1007/978-1-60327-106-6

Preface

Just as the twentieth century witnessed developments both in electronic engineering and molecular biology which have revolutionized the way we live, so the twenty-first century has been predicted to see the ever-increasing blurring of the line between them. Since so many biochemical procedures occur at the molecular level, microelectronic engineering offers the opportunity to reduce the way in which such procedures are performed to the same level. The advantages of miniaturized analysis are manifold; reducing the sample volume increases reaction speed and detector sensitivity whilst reducing sample and reagent requirements and device cost. Some of the world's largest technology companies are already involved in the development of so-called laboratories on a chip, and the field is set for rapid expansion in the next decades. The market is vast – having potential to provide, for example, bench-top versions of large and expensive equipment that could make analyses like flow cytometry as commonly available as gel electrophoresis is now.

The aim of this book is to provide biochemists, molecular biologists, pharmacologists and others with a working understanding of the methods underlying microengineering and the means by which such methods can be used for a range of analytical techniques. It describes the methods by which microengineered devices can be built to perform a number of applications and considers how the field may progress by examining some more complex lab on a chip devices which have great potential in the advancement of the way in which molecular biology is performed. We also hope that this book will be of use to microengineers, both as a reference guide for practical microengineering techniques and as a route into the development of new devices for biological applications. The union of molecular biology and microelectronics offers huge promise but one which will be all the stronger wherein both sides understand the needs of the other.

Guildford, UK
27 April 2009

Michael P. Hughes
Kai F. Hoettges

Contents

Contributors

GIUSEPPE BENAZZI • *School of Electronics and Computer Science, University of Southampton, Highfield, Southampton, UK*

KARLA D. BUSTAMANTE VALLES • *Orthopaedic & Rehabilitation Engineering Center, The Medical College of Wisconsin & Marquette University, Milwaukee, WI, USA*

DOUGLAS CHINN • *Sandia National Laboratories, Albuquerque, NM, USA*

DAVID R. S. CUMMING • *Department of Electronic and Electrical Engineering, University of Glasgow, Glasgow, UK*

SHADY GAWAD • *Swiss Federal Institute of Technology, Lausanne, Switzerland*

PAUL A. HAMMOND • *Department of Electronic and Electrical Engineering, University of Glasgow, Glasgow, UK*

STEPHEN J. HASWELL • *Department of Chemistry, University of Hull, Hull, UK*

KAI F. HOETTGES • *Centre for Biomedical Engineering, University of Surrey, Guildford, Surrey, UK*

DAVID HOLMES • *School of Electronics and Computer Science, University of Southampton, Highfield, Southampton, UK*

KATERINA KOPECKA • *Department of Physics, University of Ottawa, Ottawa, ON, Canada*

FATIMA H. LABEED • *University of Surrey, Guildford, Surrey, UK*

HYWEL MORGAN • *School of Electronics and Computer Science, University of Southampton, Highfield, Southampton, UK*

DONG QIN • *University of Washington, Seattle, WA, USA*

PHILIPPE RENAUD • *Swiss Federal Institute of Technology, Lausanne, Switzerland*

GARY W. SLATER • *Department of Physics, University of Ottawa, Ottawa, ON, Canada*

FRÉDÉRIC TESSIER • *Department of Physics, University of Ottawa, Ottawa, ON, Canada*

LEI WANG • *Department of Electronic and Electrical Engineering, University of Glasgow, Glasgow, UK*

PAUL WATTS • *Department of Chemistry, University of Hull, Hull, UK*

GEORGE M. WHITESIDES • *Department of Chemistry and Chemical Biology, Harvard University, Cambridge, MA, USA*

DANIEL B. WOLFE • *Department of Chemistry and Chemical Biology, Harvard University, Cambridge, MA, USA*

Chapter 1

Microfabrication Techniques for Biologists: A Primer on Building Micromachines

Douglas Chinn

Abstract

In this chapter we review the fundamental techniques and processes underlying the fabrication of devices on the micron scale (referred to as "microfabrication"). Principles laid down in the 1950s now form the basis of the semiconductor manufacturing industry; these principles are easily adaptable to the production of devices for biotechnological processing and analysis.

Key words: Fabrication, photolithography, photomask, etching, thin films.

1. Introduction

Fabrication of electronic devices on a micron scale was developed by the integrated circuit industry, beginning with the invention of the integrated circuit (IC) simultaneously by Jack Kilby of Texas Instruments and Robert Noyce of Fairchild in 1958. Micromachining began with the challenge issued by Professor Feynman in 1959 to build a tiny motor. In 1965, Gordon Moore of Intel postulated what is now known as Moore's Law, where the data density of integrated circuits doubles every 18 months. Today, 2006, the largest companies are fabricating complex state-of-the-art chips with over 40 photomask layers, including 10 metal layers, and are approaching 10^9 transistors in an area the size of a postage stamp. Lateral dimensions are below 100 nm and shrinking all the time. Vertical dimensions for the thinnest oxides are a few monolayers, with films of metal and dielectrics on the order of a few nanometers (nm) to a micrometer (micron, μm). A modern silicon integrated circuit fabrication facility costs over $2 billion to build

M.P. Hughes, K.F. Hoettges (eds.), *Microengineering in Biotechnology,* Methods in Molecular Biology 583, DOI 10.1007/978-1-60327-106-6_1, © Humana Press, a part of Springer Science+Business Media, LLC 2010 and equip. The new MESA micromachine laboratory at Sandia National Laboratories costs almost $500 million. Only large corporations and governments can make this kind of capital investment. Chips must be made in huge quantities with exceptional quality control to achieve an adequate return on investment. Because chip-making processes advance so quickly, process equipment is typically depreciated after 2 years.

An integrated circuit is fabricated using only four steps: film growth and deposition, patterning, etching, and annealing. A typical process begins by depositing a film, spinning photoresist, patterning and developing the resist, etching the thin film, and finally stripping the resist. By repeating these steps over and over, the most complex devices can be made. This is a gross oversimplification but it demonstrates how a number of simple steps can be combined to make very complex devices. In industry, complex processing tools such as etchers and thin film deposition systems are dedicated to a single process. Engineers in semiconductor and micromachine factories devote entire careers to reducing variability by characterizing processes and equipment and thus improving device yield. Even so, scrap rates run up to 50% of all material started in the line, depending on the maturity of the process. Mature lines with well-characterized processes will skip testing in wafer form and package every device made, rejecting the defective parts at that stage, indicating that well over 99% of all wafers were processed correctly.

Most modern wafer processing machine tools are "cassette to cassette," where all wafers are carried around the facility in specially designed wafer holders known as cassettes or boats. Humans never handle wafers. Instead, cassettes are placed into "indexers" that automatically move the wafers into the process chamber, afterward returning them to a cassette when the process is finished. Since humans make mistakes and add defects, they must be removed as far as possible from the wafers.

Here we must distinguish between the terms "fab" and "lab." Fab refers to a large, commercial factory set up to turn a profit, which produces a small variety of integrated circuit devices that have similar processes. Lab refers to a smaller facility, usually at a university or government laboratory, set up to produce a wide variety of electronic, optical, and mechanical devices with many different processes. Machinery (machine tools or just tools) in a fab is designed for high throughput (low cost) and must be designed to minimize the number of particles and defects they add to wafers. Uptime is a major consideration, and fabs will schedule routine maintenance into the process. Although programmers who use chips to create amazing programs get all the glory, the people who design and build the machines that make the chips are the real heroes because the semiconductor industry arguably has the highest technology machine tools in existence. Creating a 25-mm square chip with dozens of layers, thousands of process steps, geometries below 0.25 μm, and one billion transistors with no defects is not an easy task!

The emerging micromachine industry has benefited from all the advances made by the semiconductor industry, but is several Moore's law generations behind it. In many micromachining applications, lateral dimensions are above 1 μm, and vertical dimensions are above 100 nm. Universities and other small labs do not need the expensive dedicated cassette to cassette machines designed for high throughput and quality control. This chapter is written for the small lab, with contrasts to large production facilities. We hope to give rules of thumb that help get devices fabricated without the reader having to "re-invent the wheel" while learning to build microdevices.

Why is this chapter called "Microfabrication for Biologists" when it has absolutely no reference to biology in it? Because most commercially successful micromachines have biological applications and most of those are microfluidic devices that sense some kind of biological agent. Since a market exists, biologists have been driving micromachining away from silicon-based devices that often have moving parts to devices that analyze fluids built on glass and polymer substrates.

Many good books on the subject of microfabrication have been written (see **Suggested Reading List**). These books list some exotic and complex processes and the reader is referred to these books and others for rigid analysis and mathematical detail. Specific details can be found in the technical literature. A great deal of information is also available on the Web. A person building their first device needs practical information to get started on a project. Here we attempt to supply the reader with a lot of practical knowledge, gained from the author's experience in building devices in industry, university, and government laboratories. The approach used in this chapter is to teach a novice engineer, scientist or biologist how to design a device pattern using a CAD (computer-aided design) program and how to develop a process run sheet to get it fabricated with the help and assistance of professionals who work in microdevice process labs every day. We have avoided the use of trade names where possible. This chapter is intended to be a practical guide to getting started in the field, making no attempt to define the underlying physics and chemistry. Since most biologically oriented micromachining involves substrates that are not silicon or period II–VI or III–V semiconductor compounds, this chapter is written with that in mind.

Any device built by micromachine techniques has hundreds of potential variables to control. Small changes in any one of these can have a large impact on the final product. The keys to successful microdevice fabrication are good process control in well-characterized machines, a well-defined fabrication process, well-designed photomasks, and careful fabrication. To maximize the possibility of success, it is highly recommended that the novice microfabricator devote a great deal of time to designing the device and process before beginning production. Properly designed photomasks and a complete process run sheet, or *traveler*, can be assembled by talking with people familiar with the process tools available in the laboratory. This simply a list of all the processes and their many variables in the proper order, and a simple example is given below. In the world of microfabrication, the *tiniest* of details can determine the success or failure of a project, so a concise record of all the processing is necessary to repeat a successful process.

It is this author's experience that new scientists, biologists, and engineers building their first microdevice typically expect the first attempt at fabrication to work. This is rarely the case, so one should expect several scrap wafers before having success. The reason is that in a research setting, well-characterized, reproducible, standardized processes are not often available, and as Murphy's law states, anything that can go wrong will. With hundreds of variables to control, and only limited knowledge of any given process step, fabricating a device correctly during the first attempt is not highly probable. An engineer or biologist must fabricate the device under the limitations of the laboratory equipment available and the time available to characterize processes, but the resources available to the researcher limit processing capabilities.

2. Substrates

The first consideration for microfabrication is the *substrate*, or *wafer*, the device will be built upon. Integrated circuits are built on wafers, where micromachines can be built on a variety of substrate materials, so the term substrate is used interchangeably with wafer. Micromachining is often divided into two categories, bulk and surface. In bulk micromachining, the substrate is formed into the finished device, whereas in surface micromachining layers are added on top (or on the bottom) of the substrate, and the added layers are formed into the desired device. Many technologies combine surface and bulk micromachining. It is highly recommended that one obtain many more substrates than are required for the project to use as dummy wafers, test wafers and to take into account scrapped wafers caused by misprocessing. Plan on dropping and breaking some substrates too. It is also recommended that substrates be scribed at the beginning of the process to make traceability easier. The variables affecting the specification of substrates can be seen in **Fig.1.1.**

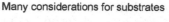

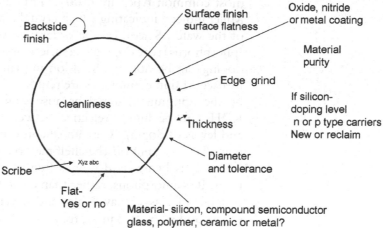

Fig. 1.1. The selection and specifications of the substrate to be used are critical variables that can determine how successful the device fabrication process will be.

The selection of a substrate determines much of the subsequent processing. Most microfabrication tools are designed for silicon wafers with a flat. The flat is simply an area ground off of the edge of the round wafer. It is there so that the crystal orientation can be determined and is used by automated processing equipment to orient the wafer. Square or other shaped substrates can be used, but non-circular substrates are much harder to work with and do not fit in many machines. Round substrates spin much better than squares do. Most laboratories have one or two standard sizes that fit into most machines, so it is advantageous to use a standard size. Two-inch (50 mm), three-inch (75 mm), four-inch (100 mm), and six-inch (150 mm) substrates are common standards. However, the metric and English sizes are not necessarily compatible in some machine wafer holders (1 inch=25.4 mm).

When ordering substrates, several variables must be accounted for. After diameter, and thickness, one must determine the flatness, surface roughness, edge grind, the finish on the back of the substrate, temperature limitations, coefficient of thermal expansion, and transparency to a given wavelength of light and mechanical toughness. Poorly ground edges can be a source of particles when they chip, and can initiate cracks which will cause the substrate to break. In a plasma etch process, note that the conductivity of the substrate can have an impact on the way the material etches.

Silicon has the advantage that it is widely available, relatively inexpensive and very well characterized. It is also the most perfect material obtainable, being a slice of a single crystal. The crystallinity can be taken advantage of in directional etches and in applications where it is desired to etch through the wafer. The most common types are (100), (111) and (110). These are the Miller indices, indicating which crystal plane makes up the surface of the wafer. Silicon can be ordered with either p (boron) or n (phosphorus) type doping in a variety of conductivities. Additional doping can be done with an ion implanter or high-temperature furnace. If thin membranes are required in the device, silicon may be the optimum material because some wet etchants, such as KOH, can be highly selective between different crystal planes and levels of doping. Custom silicon can be ordered from many manufacturers, but off-the-shelf material is usually much cheaper.

Glass is being used more often in micromachining applications. It is amorphous, rather than crystalline like silicon. Dozens of types of glass are available, and it can be ordered in custom shapes and sizes. It is inert, transparent, and mechanically tough. Some types of glass can be very inexpensive, such as microscope slides, and some, such as fused silica and quartz, can be very expensive. Many glasses are chemically impure, and because it is amorphous, it can have a non-uniform microstructure. This can affect etching, optical, and electrical properties.

Polymer substrates are now being used. It is very difficult to generalize about polymers, since so many are available. Plastics are inert to acids and can be formed by molding and embossing, but plastics tend to have a non-uniform microstructure. Polymer substrates can absorb water and solvents, changing their dimensions.

Metal substrates have been used, but due to the reactivity of metals they generally hold up very poorly during chemical treatments. It is also difficult to get surfaces with low roughness in metals. *Ceramics* are also used, but the polycrystalline nature of ceramics can present severe processing difficulties

Wafer supply companies often have film deposition equipment. If all of your substrates need a particular film, it may be cheaper, faster, and cleaner to buy wafers with a film already deposited on them by the supplier.

3. Photomask Design

The objective of a photomask is to selectively block part of the light used to expose some kind of photoresist. The most common kind is made of a thin film of chrome on a square of glass that is transparent into the UV range. Many different kinds of glass are available. The chrome is often oxidized to a gold color to cut down on stray reflectance from the front surface. Most align tools use a mask that is 1 inch/25 mm larger than the wafer size.

Contact the commercial mask fabrication shop at the beginning of the design cycle so that your design is compatible with the shop's tooling. Many drawing programs are available, but line types and shapes used by generic drawing programs cannot be recognized by mask shop tools. Mask shops require the data to be in specific formats, so an important factor is being able to convert the finished CAD drawing into a format that the mask shop can use to produce a mask. Any software for IC design work is capable of saving the data in an acceptable format.

Inexpensive photomasks can be designed using any drawing program and printed out using standard office printers on Mylar transparencies. Such masks are only effective for larger geometries with loose tolerances, but can be made quickly and cheaply.

A *reticle* is a photomask that is scaled, typically used in various types of steppers. The advantage of using a reticle over a photomask is that small geometries are achieved optically by the end user, rather than by trying to write small features directly on the photomask. Each type of stepper requires its own reticle, where photomasks can be used in most machines, if sized right. If your device is small and several copies can fit on a single wafer, a shop may generate a reticle and step it over and over to make a mask, rather than write the same pattern over and over.

Sometimes a photomask has a *pellicle* attached to it, a thin membrane that is set away from the chrome geometry by a spacer. It keeps particles away from the mask surface and thus out of focus.

3.1. Photomasks

Many designers have been frustrated because an expensive mask does not do what it was designed to do. As illustrated below, the mask design is not independent of the process. Several decisions must be made during the design stage.

1. Wafer or substrate and die size
2. Type and size of mask
3. Dark or light field
4. Spot size
5. Minimum dimension/pitch
6. Defect density
7. Tolerances
8. Process bias
9. Positive or negative resist
10. Alignment keys
11. Handedness
12. Dicing the wafer into individual chips
13. How the device will be packaged or how the inputs and outputs will travel from the macroworld to the chip scale microworld.

One of the first decisions to make is the *die*, or *chip*, size. The word die comes from the IC industry, where a typical wafer will have many dies on it. When you purchase a microchip in a package or see one soldered into a printed circuit board, there is a small silicon (or compound semiconductor) chip inside that could be anywhere from a fraction of a mm^2 to ~30 mm^2 in size. These individual dies are cut from a larger wafer. Many micromachine devices can take up an entire wafer. If your device is smaller than the wafer it may be beneficial to have many dies on the same wafer.

The type of lithography tool to be used and the substrate size determine the *size* and *type* of the mask to be made. Contact printers, steppers, and projection printers all have many specific mask requirements, so decide on the alignment tool before doing the CAD design. Masks can be ordered as masters or copies. Copies are cheaper than masters, but can have lower resolution. A master must be made to make a copy. If contact printing is to be used, a mask can be damaged, so it may be economical to use copies rather than the original master.

Dark field masks are mostly chrome with holes in the film; *light field* masks appear to be mostly glass with chrome geometries on them. The type of resist used and the desired pattern determine whether dark or light field is called for. Masks are usually drawn light field, and dark field is specified on the mask shop's run sheet. The way a mask exposes photoresist is a function of how much geometry is present and how it is distributed. During lithography, dense geometries in one area may expose differently than areas with low density geometry. If a plasma etch is to be used, etch rates are a function of geometry density and size.

A mask shop starts with a photoresist covered chrome film and writes a pattern using a laser or an electron beam. The *spot size* used for the writing is a major determinate of the price, with 0.25 μm being common. Smaller spot size gives smaller geometry, but takes longer to write. Both line and space (pitch) *minimum dimensions* on the mask can determine what type of equipment is used to make the mask: larger dimensions are sometimes cheaper to make. The smallest geometry can drive the price of a mask. Another factor to consider is *pitch*, the sum of the line space. Mask shops may define a minimum pitch.

Another factor determining photomask price is the number and size of *defects* allowed. If your device has small, dense geometries, very few defects in the form of missing chrome or extra spots of chrome can cause fatal defects in the device. If you have large, low density geometries, small defects may not affect the device at all. As a general rule, a defect 10% of the geometry size may cause problems.

The *tolerances* in a given layer and the tolerances between layers can also impact the price of a photomask. If all lines and spaces are above 1 or 2 μm the mask can be fairly inexpensive. Mask shops cannot alter their quality control for each mask. If a mask has tight tolerances and low defect levels, a shop may have to make it several times to get it right, increasing the cost. Masks often have special geometries specifically for critical dimension (CD) measurements. Understand that any metrology tool will have error in its measurement and that measurements can vary across a wafer. Ideally measure a CD at several points on a wafer or mask and average the numbers to understand how much error is in a process.

A very critical idea that must be incorporated into any mask design is *process bias* (**Fig. 1.2**). The geometry dimensions laid out

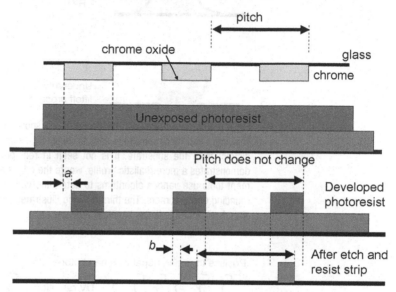

Fig. 1.2. Process bias during exposure and etching. Upon exposure, the photoresist will change from the mask dimension by some factor 2*a*. Depending upon processing conditions, *a* can be positive or negative relative to the mask dimension. After etching, another bias is introduced, 2*b*. Etching conditions and the resist profile determine *b*. The pitch is constant throughout all processing. The designer must account for these biases when laying out the photomask.

in the CAD work of the photomask are very unlikely to be reproduced exactly on the finished product. Every time a pattern is transferred it will change dimensions, growing or shrinking, and corners will be rounded (**Fig. 1.3**)

With *positive* tone resists, the part that is exposed becomes soluble in the developer and goes away after develop; with *negative* tone resists, the exposed part becomes cross linked, and the unexposed parts are removed during develop (**Fig. 1.4**). Typically with negative resist, openings in the chrome translate to larger geometries after exposure. Positive resists re-create the photomask, but chrome patterns will shrink after exposure and develop.

Most processing today is done with positive photoresists. Positive resists give finer geometries than negative resists. Because positive resist developer chemistry uses aqueous bases it presents less waste disposal issues than negative resists which use solvent-based developers and rinses.

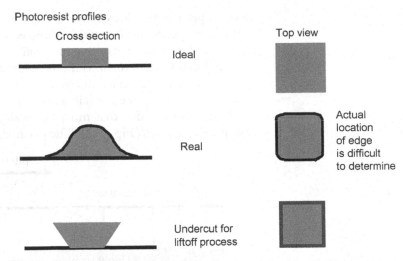

Fig. 1.3. Photoresist profiles. The ideal profile is shown. Its transfer characteristic will be very good and its dimension is easy to measure, since its edges are well defined and distinct from the substrate. It is not seen in real processing. The second drawing demonstrates a more realistic profile, where the edges are not distinct at all. Measurement tools use various algorithms to determine where the edge actually is. Note the rounding of the corners. The third drawing illustrates what is possible in resist processing. Such a profile is used in liftoff processes.

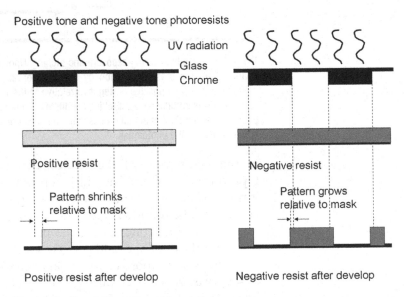

Fig. 1.4. Positive and negative photoresists.

Well-designed *alignment keys* are critical mask features for devices that have more than one layer, since the second layer must be aligned to the first layer.

Many different alignment keys have been used, and each designer has a favorite. Understand what works well in the lithography tool before doing the design. In the figure, both light and dark field keys are shown as crosses. This author likes to use crosses about 100 µm long and 20 µm wide for the first layer for contact printing. The layer to be aligned to it has a cross 98 µm long and 18 µm wide, giving a tolerance of about ± 1 µm. Thus, a smaller cross fits on top of a larger cross that is etched into the wafer. Recall that the mask design dimensions and the geometry dimensions on the first layer will vary based on the process biases introduced, so those biases need to be understood before laying out the align keys. Most often all layers are aligned to the first layer, so the first layer needs keys for each layer to be aligned to it. Align keys must be positioned on the mask and die so that the microscopes on the align tool can easily locate them.

A very common mistake in mask layout occurs in dark field masks. Trying to find align keys using a microscope on top of a dark field mask can be very difficult. Since most of the mask is chrome, the underlying layer can only be seen through the align key itself. Draw a large box of clear geometry and put a small chrome align key inside the clear box. The large box makes it easy to find the align key and makes it much easier to find the align key on the underlying layer (**Fig. 1.5**).

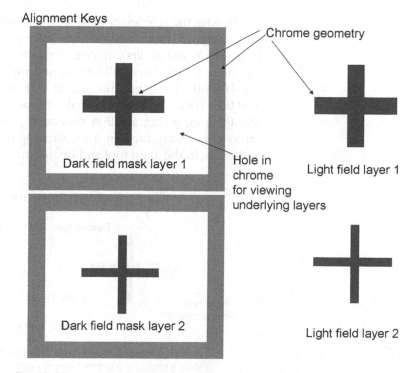

Fig. 1.5. Alignment keys for dark- and light-field masks. On the left, align keys in a dark field mask must be surrounded with a large clear area to make it possible to see the geometry underneath that a layer is being aligned to.

Another common mistake is to draw closed shapes by putting one square or circle on top of another (**Fig. 1.6**). Human eyes can see the resulting donut, but a computer generating a mask sees two circles on top of each other and writes both. What was supposed to be a donut or outline becomes a solid shape. Therefore, closed shapes must be drawn as two parts, as seen in **Fig. 1.6**.

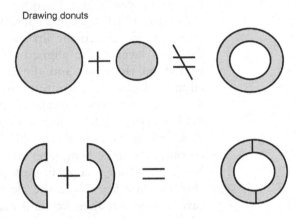

Fig. 1.6. Drawing closed shapes. A computer recognizes only solid shapes and cannot subtract one pattern from another.

Besides the tolerances of specific geometries, another important tolerance indicates how each layer lines up to the previous layer. Masks and wafers can have *run-in* and *run-out* errors. These are where a design may call for a dimension across an entire wafer, say 100.000 mm, but the actual dimension is, for example, 100.005 mm. The layer placed on top of that layer may be 99.009 mm, giving a 0.006 mm or 6 μm run-out error. Such errors can be introduced at the mask shop by bowing of substrates and masks in the align tool and by thermal mismatch (**Fig. 1.7**).

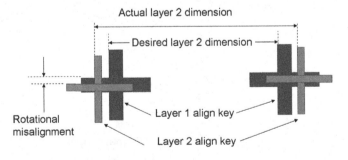

Run out error upon aligning layer 2 to layer 1. The left align keys show a rotational alignment error as well.

Fig. 1.7. Alignment and runout errors seen upon alignment of the top layer (*light gray crosses*) to the layer etched into the wafer (*dark crosses*).

Taking *handedness* or chirality into account is important when designing a photomask. Masks are usually used chrome side down or viewed through the back of the mask relative to the design artwork. One can buy boxes of unexposed photomasks covered with chrome and photoresist, quite useful in the typical laboratory. Do not open them except in yellow lights. By placing an existing mask in contact with a blank mask, a copy can be made. Typically some pattern fidelity is lost, and the copied mask is a left-hand version of the original. Copying a mask a second time restores the chirality to its original. To reverse a mask, i.e., to use it with negative photoresist, for example, one can make a copy, strip the resist, sputter or evaporate a different metal onto it (such as titanium), etch the chrome in a chrome specific etchant, and have a titanium reversal copy. Edges may be somewhat rougher than the original.

Sometimes test patterns that can define resolution, or measure overlay alignment (such as a vernier pattern), are added to the photomask. Many chip designers have various test patterns available to cut and paste into designs. Electrical devices often have special test devices such as resistors or individual transistors built into a device in the test pattern section of the chip.

If multiple devices, or dies, are on a single wafer, it is usually *diced* up as a final processing step. The blades used in dicing saws are as thin as 200 μm, but by leaving a scribe line between dies of at least 500 μm makes the dicing operation much easier.

Packaging is often the most expensive part of an IC and is often overlooked during the device design stage. Final packaging can be a very difficult part of building a device. In ICs, only electrons go in, and only electrons and heat leave the chip. In micromachines, electrons, heat, fluids, mechanical energy, photons, and perhaps even magnetic energy all may go in and out; consequently, the package can be very complex and probably cannot be purchased off the shelf. It is difficult to get information from a micro level to the human who will make decisions based on what happens at the micro level. Thus the designer of micromachines must take the package into account early in the design process.

Consider when designing a mask how the finished device will be contacted electrically. Electrical input and output to the outside world is usually done with *bond pads*. Bond pads are squares of metal that are the terminus of electrical wires patterned on the chip. Special micromanipulators are available to put tiny probes onto bond pads, but if space allows, bond pads 2 mm on a side can be contacted with the crudest probe tips. We recommend making them as large as space permits.

Another type of mask is a *shadow mask*. This is usually a thin sheet of metal with holes cut in it. The shadow mask is placed in contact with the substrate and put into an evaporator or sputterer (below). Metal is deposited through the mask, eliminating the need for lithography. It is cheap and quick, but edge definition is poor and it is difficult to align shadow masks to existing patterns. Resolution is limited to above 50 μm. Shadow masks can be made in a machine shop, or even by putting pieces of tape on top of the device. Shadow masks can be ordered commercially from companies that do lithography on stainless steel sheets and etch patterns into them.

---

4. Lithography

To build a micromachine or microelectronic device, many variables must be controlled. Below we outline the issues.

4.1. Spinning

The *adhesion* of spin-cast films to the substrate is an important variable, and must be dealt with before spinning a film. Modern resists adhere well to many materials, and dry (plasma etch) processes usually do not affect the adhesion. However, when subsequent processing involves mechanical stress or liquid chemicals, the interface between the spun film and the substrate may be attacked. The most common and well-proven adhesion promoter (primer) for photoresist is HMDS (hexamethyldisilazane, $(CH_3)_3SiNHSi(CH_3)_3$). This material is designed for silicon-based materials and may not work on metal films and polymers because it is designed to work with hydrated surfaces. Its reaction, $R\text{-}OH + (CH_3)_6Si_2NH = 2\ R\text{-}O\text{-}Si(CH_3)_3 + NH_3$. R is a substrate molecule such as silicon. Ammonia is liberated from the reaction. Primed wafers do not need to be baked, but a 124°C bake can help eliminate the ammonia. HMDS makes a previously hydrophilic surface appear to an organic molecule as though it is a hydrophobic organic surface. It is often used as a diluted solution (~20% in an organic solvent) spun on at 3,000 rpm prior to spinning the photoresist. More modern labs will use a vapor prime system, where the adhesion promoter is released in vapor form into an evacuated chamber, usually well above room temperature. Vapor prime systems use less material and coat very uniformly. The key to using adhesion promoters is that they must be a monolayer, forming a bridge between two incompatible surfaces, such as an oxide and an organic. Adhesion promoters designed to work on many classes of material are available, so research an adhesion promoter that forms a good interface between the two materials you are using if HMDS is not effective in your application.

By dispensing a small amount of a liquid-phase material onto a wafer and spinning it around at high speeds, very uniform films can be put on any substrate. Commercially available materials designed for spinning may have several solvents in them to control evaporation rates and film uniformity. Often in a research lab new and novel materials need to be spun, so the engineer must first characterize the material's spinning characteristics before devoting the material to a device. A very simple spinner can be made using a small motor and a chuck appropriate to hold the substrate. However, most labs have commercially available spinners with vacuum chucks. When working with odd-sized substrates, a sheet of red silicon rubber with a hole in the middle can be placed on most spinner chucks, allowing the chuck to successfully hold the substrate. Always close the lid of a spinner to control air flow around the spinning wafer, keep solvent vapors in check and to protect you from objects thrown off the chuck.

The variables of interest when spinning are as follows:

1. solution viscosity and solids content

2. solvent system vapor pressure

3. acceleration rate

4. final spin speed

Other factors can determine the thickness and quality of the film, such as the atmosphere inside the spinner and the air flow around the substrate while spinning, but these variables are secondary. With commercially available materials, an amount of solution is poured into the center of the substrate roughly one-fourth the diameter of the substrate. With lab-made solutions, the quantity may need to cover the entire wafer to get good coverage. Solids content is important because films will tend to crack if the volume changes too much when the solvent evaporates. Spin speeds below 500 rpm rarely work well. A 3,000 rpm spin with at least 1,000 rpm/s up to 5,000 rpm/s acceleration rate is a common place to start when characterizing a material on a spinner. Thirty seconds to one minute is usually an adequate spin time. High vapor pressure/low boiling solvents can evaporate so quickly that a film forms on top of the liquid creating a poor spin. Low vapor pressure/high boiling solvents can take so long to evaporate that film dries like a "coffee stain" leaving a poor-quality non-uniform film, so selection of solvents is an important factor when designing materials for spin casting. Particles on the surface of the wafer or in the resist can create "comet tails," a common spin-cast film defect. These are summarized in **Fig. 1.8.**

Photoresists have a short shelf life, but often work well a few years beyond that date if stored in a refrigerator. Production fabs never use any chemical past its expiration date. To prevent contamination, always pour resist, or any laboratory chemical, from the factory bottle into a dry, cleaned brown bottle for use. Never dip any pipette or other items into the clean bottle of resist. Keep bottles closed and never expose resists to room light. It is a good idea to wrap photoresist bottles in aluminum foil to keep light out.

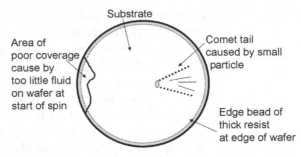

Common defects seen when spinning photoresist

Fig. 1.8. Photoresist spinning defects. The edge bead is normal and usually causes no problems. It is sometimes stripped off by carefully spraying a stripper on the edge of the wafer while it is spinning.

After casting, photoresists need a *soft bake* to dry out the residual solvents. Improperly dried resists can stick to the photomask or develop poorly. Baking can be done on a hot plate or in an oven, near 100°C, depending on the resist type. Bake times on hot plates are typically 1–2 min, oven times are usually 10–30 min. More uniform temperatures are available in ovens. Some more sophisticated labs have the bake plate in line with the spinner, and industrial fabs have the spinner, bake plates, and develop systems in line with the stepper for a fully automated process. In such highly automated process equipment it can be difficult to change parameters for experiments or to use non-standard substrates.

Clean the spinner and replace the liner before spinning your wafer. Vacuum out lint left over from the wipers used in the cleaning procedure. Always spin a dummy wafer with resist before committing the actual device to gather particles left over from cleaning and verify that the spinner is working properly. These dummies can be used to characterize the lithography process, so store them under yellow lights.

Spin on glasses (SOG) are solutions containing chemicals that can be thermally degraded into oxides and can be spun onto wafers. Any spun on material has an advantage over vapor deposited materials because they planarize, or level, the surface, analogous to snow falling on a field or water filling a shallow pool.

Some materials designed to become part of the microstructure can be made photosensitive, such as SU-8 (a trademark) negative resist and polyimides. Both families of materials are very tough in their cured form and can be very simple to process. These kinds of polymeric materials often require thermal treatments to cure them, so heat cycles must be accounted for in the overall process.

4.2. Lithography Tools and Exposure

Most lithographic patterning is done using a mercury or mercury argon light source, which has peaks at 365 nm (i-line), 408 nm, and 436 nm (g-line). Some machines filter out specific wavelengths to improve resolution. The optimum time to expose a photoresist depends on the absolute wavelength, the relative intensities of the peaks, resist sensitivity, resist thickness, desired profile, and the exposure tool. The reflectivity of the substrate can also affect the exposure time and the quality of the image. UV light sources age, therefore total intensity and relative peak intensity can change over time in any given exposure tool. Most labs have a meter for measuring the light output to aid in setting the exposure time.

Several types of tools exist to transfer a pattern from a photomask to a substrate. *Contact printers* place the chrome side of a photomask in contact with the photoresist on the substrate. These machines are fast, reproduce the mask well, and have high reliability. The disadvantage is that they are typically limited to 1 μm and above in geometry size and they damage the photoresist and mask through physical contact. Modern versions of these tools often have backside alignment capability, where photographic or infrared methods can be used to align patterns on both sides of a substrate. IR methods only work on silicon or other IR transparent substrates that have smooth surfaces.

Proximity printing is similar to contact printing, but the mask never comes into contact with the photoresist. Resolution is lost, but neither the mask nor the resist is damaged.

Projection printers separate the mask and the substrate through complex optics and give resolution similar to that of contact printers. These were used extensively in the IC industry for many years due to speed and the quality of the image they project, but have fallen out of favor in recent years.

Steppers are the workhorse of industry and are found in larger universities and corporate labs. They are expensive to buy and maintain, require specialized photomasks, and are usually operated by specialists. Steppers can expose large 300-mm wafers with no defects added. A $10 \times$ reduction stepper can create submicron geometries by reducing a 5-μm line on a reticle to a 0.5-μm line on a wafer. Because of the size reduction, the image must be placed on a wafer over and over until the entire surface is exposed, known as step and repeat. These tools do not work well with large die sizes as are common in micromachines. If large high-resolution patterns are required, different pieces must be stitched together. Steppers require special align keys to be put onto the wafer known as globals.

The smallest geometries are still made using an *electron beam writer*, often a modified electron microscope. Geometries as small as the spot size of the e-beam can be written, but only very tiny areas can be exposed. E-beam writing is a very time-consuming technique and has not found commercial use, except in photomask generation.

Over the last few years, many researchers have had excellent results using polydimethoxysilane, or PDMS, to make "rubber stamps." A conventional photomask is designed, and the image is etched into an appropriate substrate. A PDMS film is cast onto the substrate. After peeling the polymer off of the substrate, it can be coated with a variety of materials, which are then transferred to other substrates. The reader is directed to the original literature, much of it from Professor George Whitesides' laboratory at Harvard for details on this useful technique.

Another technique is to use synchrotron X-rays to image polymethyl methacrylate substrates (PMMA) creating very high aspect ratios up to a millimeter thick. This technique was developed in Germany and has the acronym of *LIGA* (lithographie, galvanoformung, and abformung). Very high resolution can be obtained in high-aspect structures, but since time on a synchrotron is difficult to get, the technique is limited to a few laboratories worldwide.

4.3. Developing

After being exposed through a photomask with UV light, the exposed regions (positive resist) or the unexposed regions (negative resist) of photoresist must be removed. Developing can be done in a beaker or in a sophisticated spray develop spinner. Most small labs use a beaker for immersion develop. Sufficient agitation is necessary to get the dissolved photoresist away from the surface so that a fresh developer can get to the surface. Spray developers eliminate this problem, but require far more chemicals and considerable characterization of the process. The develop time depends on resist thickness and type, developer concentration, agitation rate, and temperature. For a 1-μm-thick positive resist with a 3–2% concentrated basic developer it is roughly 1 minute. Do not re-use or save positive developers because they can absorb CO_2 from the atmosphere and become less effective. Common bases are tetramethyl ammonium hydroxide (TMAH) and sodium hydroxide. Sodium is a severe contaminant in silicon processing, so TMAH is sometimes known as an MIF, or metal ion free, developer. Developers may have surfactants in them to assist in wetting.

Develop time can determine the resist profile. Overdevelop narrows CDs, underdevelop can leave residual films on the surface or leave a thin film around the edge of the pattern. Plan on doing several exposures on dummy wafers to determine the proper exposure for your combination of substrate and resist. Typically, dummy wafers are both over- and underexposed to bracket the ideal exposure time/develop time around the optimum. Use optical microscopes to inspect the resist pattern, but microscope light can expose resist. Underdeveloped resist can be re-developed to improve its appearance.

After developing, a rinse step must be done. Negative resists use organic solvents to remove the residual developer, positive resists use DI water. A minute or so of immersion is usually adequate, but flowing fluids are best. In small labs, a nitrogen blow gun is often used to dry the wafer after rinse (see below about blow guns). Hold the wafer vertically using clean tweezers with the bottom edge of the wafer on a special clean room wiper. Blow from top to bottom, doing the front and back alternatively. The cloth will wick off the last drops, which are nearly impossible to blow off into the air. Automated develop tracks spin the wafer dry, a much better alternative.

Resist patterns that are unacceptable can be stripped off and re-spun even after exposure, a process known as re-working. Often a ghost pattern can be seen on the substrate where the original pattern was.

4.4. Descum

Often after develop and soft bake, some residual photoresist can be seen around the edges of the pattern or out in the field of the wafer. This can be alleviated by doing a short oxygen plasma, often called a descum. Typically 30 s at 400 mt/100 watts in pure oxygen is affective. Such a process should remove 100–300 nm of resist. As with all other processes, characterize it before committing wafers with value, but doing a descum prior to etching is often considered to be good practice. Sometimes a descum is done before depositing the resist to improve adhesion by roughening the substrate or etching away organic films.

4.5. Hard Bake

Many resists require a "hard bake" after develop to harden the resist for further processing. This step is important in wet etching, but may make the resist difficult to remove. Hard baking is done on the same equipment as the soft bake, but is usually done about 10°C hotter, with typical hot plate times for a 1-μm -thick resist of about 2 min.

4.6. Resist Stripping

Photoresists are designed to be stripped off after transferring a pattern to a material beneath the resist layer. Positive resists are often easier to strip off than negative resists. (Photoresist stripping steps are often left out of process run sheets.) Piranha clean (detailed below) works very well to remove both positive and negative tone photoresists, but can attack underlying layers. Acetone strips positive resists that have not had much post-processing that hardens resists, such as baking, ion implant, or plasma etching. Acetone should always be followed by an isopropyl alcohol (IPA) rinse. Positive resists can also be stripped by exposing the developed pattern with a blanket exposure in the alignment tool or another UV source, and stripped with the same developer used to delineate the pattern in the first place. Acetone/IPA and re-expose/re-develop are effective in stripping resists for rework. Many commercially made photoresist strippers are made, but check with substrate compatibility before using these products.

Another method to strip photoresists is to use an oxygen plasma, sometimes known as ashing, which is just a long descum process. Ashers can be much simpler tools than the plasma etchers discussed below. Barrel etchers or small parallel plate etchers work well for this step. Usually a flow rate of oxygen or forming gas (15% hydrogen, 85% nitrogen) to achieve a pressure of 100–500 mt is used, with a power setting between 100 and 300 watts. Time depends on resist type, thickness, processing, and the etcher configuration and settings. Times can be from a few minutes to hours. Many substrates are not affected by oxygen plasmas, other than some minor sputtering, but some substrates can be etched in oxygen, particularly polymers. Even mild ashing can damage substrates and leave undesirable defects. Often a wet cleaning is done after ashing to remove residual particles. Sometimes a wet cleaning is done first to remove the majority of the resist, followed by a short plasma clean to remove residue left by the wet stripper.

It is important to do long pump downs prior to introducing the etch gas because substrates or films can outgas, changing the chemistry of the plasma. Water and other contaminants on the surface of the plasma chamber are slowly pumped away. Always note the color of the plasma as a quality-control check. Oxygen plasmas look bluish. If the system has a leak and nitrogen is present, the plasma will have a pink tint to it.

All surfaces in a typical lab will have a thin film of organic residue on them from the environment or non-volatile organic residues left by solvents. Argon plasmas using the conditions above for 30 s to 1 min or so can sputter away residues, and oxygen or CF_4 plasmas can etch them away. Such activated surfaces will return to their original condition within a few minutes of leaving the vacuum chamber. Such a treatment can be very effective for all types of substrates to improve adhesion and may be used just prior to deposition of an adhesion promoter.

4.7. Substrate Cleaning

Cleaning wafers or substrates is often overlooked or viewed as an afterthought, but doing it well can make a lot of difference in the performance and yield of the final product. Wafers may need to be cleaned several times before they are completed. New wafers should be cleaned prior to any processing.

A common and very effective method to clean silicon or glass substrates is using *piranha* clean. It is so named because it eats almost anything organic. This is a mixture of 98% concentrated sulfuric acid and 30% hydrogen peroxide above 100°C. Mixing the two chemicals is exothermic. It destroys almost any carbon-based compound and many metals, especially aluminum. The ratio of acid to peroxide is from 2:1 to 10:1. Many labs have a permanently set up bath for this process. The peroxide breaks down into water and oxygen and must be periodically replaced. To work well, the solution should look like a clear bubbly soda pop solution (lemonade in Europe, Sprite in the US). Needless to say, it is very dangerous to humans. Wear an apron, acid-resistant gloves, and a face shield when working with this material.

Hot piranha can only be put into glass or Teflon (fluoropolymer) containers. All wafer holders placed into the bath must be made of Teflon. Do not put plastic coated metal objects into a piranha bath. Typically 5 min cleans any silicon or glass surface. Metal and plastic substrates cannot be cleaned in piranha. It is highly recommended that commercially available wafer handling boats and tools be used in piranha clean baths.

Waste piranha bath can be stored in polypropylene or polyethylene containers when cool. Do not re-use solutions after removal from the original bath. Disposal is problematic because the peroxide can remain active and create gas for many weeks after the solution is cooled. Do not put waste piranha in a tightly sealed container.

A 5-min rinse in flowing deionized water removes sulfate from the surface after cleaning. This is best done in a cascade or dump rinser. Cascade rinsers flow copious amounts of pure deionized water over a weir from back to front. Place the clean wafers in the front (relatively dirtier) tank, and after 5 min transfer them to the back (cleaner) tank. Dump rinsers fill the tank, let the water out through the bottom and re-fill the tank. If no special rinse tank is available, rinse in flowing water. Particles tend to float on the surface of liquids, and any time you pull a wafer through a surface between air and liquid it can pick up the particles on the surface, analogous to a Langmuir Blodgett trough.

If a sulfuric acid based clean is too aggressive for your substrate a boiling mixture of 25% ammonium hydroxide (NH_4OH), water and hydrogen peroxide (H_2O_2) can be used instead. Use a ratio of roughly 1:5:1 for about 10 min. This is sometimes known as an RCA1 or SC-1 clean and is often done in conjunction with an ultrasonic or megasonic agitation. Ultrasonic energy can damage some substrates. This clean removes particles and organics.

Some laboratories use hydrochloric acid based cleaning solutions, such as $HCl:H_2O:H_2O_2$ in a 1:6:1 ratio, sometimes called an RCA2 or SC-2 clean. (The process was developed by the RCA Corporation, but is now known as a semiconductor clean.) This clean was designed to reduce metal contaminants on the surface, not typically applicable in micromachining applications. This solution rinses off of surfaces much better than do sulfuric acid based cleans.

Another way to clean a surface, especially silicon, is to etch the native surface oxide off in very dilute hydrofluoric acid (1%). Most metals and all silicon have an oxide on the surface. Silicon oxide or silicon dioxide is hydrophilic or water loving. After a few seconds in HF, the surface becomes hydrophobic or water hating. This is easily seen because silicon wafers "dewet," where the acid sheets off the wafer leaving an almost dry substrate when pulled out of the acid. By etching away a thin layer, particles or sulfate-laden layers left over from piranha cleaning can be removed from a surface.

Small particles (below about 5 µm) CANNOT be blown off of a surface with an air stream or flowing liquid. Fluid streams have a velocity of zero near the surface. Below about 1 µm, the adhesion force of a particle is far greater than any force reasonably possible with a moving fluid stream.

4.8. Rinsing and Drying

Rinsing and drying are perhaps the single most important step in any wet process. Dedicated wafer processing labs have clean water delivered in special piping, detailed below. Clean tanks, glassware, and other hardware are also very important for getting clean surfaces because a fingerprint inside a rinse tank can lead to an organic film on a wafer. Regardless of the rinse method used, never let water drops dry on a wafer surface. DI water is very corrosive and will leave water stains.

A very good tool for rinsing a wafer after a cleaning or wet etching process is the spin rinse dryer. After an immersion rinse for 5 min in flowing DI water, substrates in a boat are spun while being rinsed with more DI water. A final high-speed spin with flowing filtered nitrogen finishes the process and leaves clean, dry wafers. Since entropy always increases, a freshly cleaned wafer surface is a magnet for particles. Immediately place the clean wafers in a clean container. As with all tools, verify that the machine is clean by running clean dummies in it before processing your wafers.

More modern facilities use variations on alcohol dryers. These tools are based on the Marangoni effect, which is defined in the IUPAC Compendium of Chemical Technology, 2nd ed. 1997, as "Motions of the surface of a liquid are coupled with those of the subsurface fluid or fluids, so that movements of the liquid normally produce stresses in the surface and vice versa. The movement of the surface and of the entrained fluid(s) caused by the surface tension gradients is called the Marangoni effect." These dryers can leave very clean dry surfaces and lower a lab's water usage. However, they are larger tools and not often found in smaller laboratories, but are the ideal tools for drying wafers. Effectively, alcohol displaces the water on the surface of a wafer and surface tension effects remove particles.

5. Thin Film Deposition

5.1. Vacuums

At the heart of most types of microdevice fabrication are vacuum systems. Thin films of metals and oxides and silicides can be deposited using vacuum systems. Many materials can be etched in vacuums, thus it is instructive to review some basics of vacuum systems. The somewhat antiquated unit of torr is still in use in vacuum systems. One atmosphere $= 760$ torr $= 101$ kPa, one torr $= 133.322$ Pa, 1 Pa $= 1$ newton$/m^2$.

Pumps can be categorized by the basic pressure they reach, which roughly corresponds to the three regions of pressure: low vacuum (atmosphere to 10^{-3} torr $= 1$ mt), high vacuum (10^{-3} to 10^{-7} torr), and ultrahigh vacuum, (10^{-7} torr and below). Most wafer process tools operate at base pressures in the high-vacuum region.

Leaks into vacuum systems from cooling systems, bad fittings and feedthroughs are common problems. Another common issue is contamination in the chamber. Fingerprints and other surface contaminants can be "virtual leaks." A fingerprint on the inside of a clean vacuum system can outgas and coat the entire chamber and your wafer, with a thin film of oil, making it very difficult to reproduce some results. Another type of virtual leak is from air or gas trapped underneath a screw in a blind hole, or gas trapped inside a microstructure. Films, particularly polymer films, can also outgas and slow down pumping times. Every time a vacuum chamber is opened the walls absorb water and other atmospheric contaminants which must be pumped away when the system is closed up. All systems leak, due to outgassing and poor seals. The simplest way to leak check a system is to pump it down, close all valves to all pumps and watch how fast the pressure rises. If it rises faster than what was historically normal for the system, a leak may be present. Systems need to be opened and manually cleaned on occasion, depending on their use.

Two pumping regimes are defined. The first is the viscous flow regime, where a gas behaves as a fluid and the mean free path of a molecule is much much smaller than the chamber size. In the molecular flow regime, molecules do not interact with each other and the mean free path of a molecule bouncing around inside the chamber is much greater than the size of the chamber.

There are several types of vacuum pumps, each with its advantages and disadvantages.

The most basic and common type of vacuum pump is the *rough pump*. These pumps are designed to move a lot of gas quickly from atmospheric pressure to about 50 mt, roughly where viscous flow ends. Some simple plasma etch systems have only a roughing pump. These pumps have a rotating shaft and chamber immersed in oil that moves gas molecules from the low-pressure side to the high-pressure side. When used for pressures below about 100 mt, any pump with oil in it will have some backstreaming of oil into the vacuum chamber. This can be minimized with the use of a trap that absorbs the oil. The trap requires routine cleaning and maintenance. Two types of lubricants are used in roughing pumps – hydrocarbon and fluorinated oils. Hydrocarbon oils are cheap, but cannot be used in systems that pump oxygen or other reactive gases. Fluorocarbon oils are very expensive, but are also very inert. It is our experience that pump oils are rarely changed in university laboratories. Changing the oil can improve pumping capabilities, reduce chamber contamination, and increase pump life. Treat used pump oil like toxic waste. Modern systems use more expensive oil-free dry roughing pumps.

High-vacuum pumps come in three common types, the *oil diffusion pump*, the *cryogenic pump* and the *turbomolecular pump*. All high-vacuum pumps operate in the molecular flow region, so these pumps all rely on gas molecules randomly moving to the pump. All high-vacuum pumps take a system from ~50 mt to the high-vacuum regime, so the majority of the gas in the chamber must be removed with a roughing pump prior to opening the valve that exposes the chamber to the high-vacuum pump. Oil diffusion pumps have been around for many years, are cheap and work very well. They take a long time to heat up when cold and have limited pumping capacity. Sometimes these pumps burn the oil and must be cleaned. Oil diffusion pumps can reach pressures less than 5×10^{-8} torr. Cryogenic pumps or cryos are very expensive and remove gas from a chamber by freezing it out on a material that is held near the temperature of liquid helium, around 20 K above absolute zero. They are very "clean" pumps since they add no contamination whatsoever, but some cannot pump out helium very well. On a routine basis cryogenic pumps must be regenerated, where the cold head is warmed up so the frozen gases can evaporate. This takes several hours and can result in considerable system downtime. Cryo pumps can achieve pressures as low as 5×10^{-9} torr. Turbomolecular or turbo pumps have increased in popularity recently because of improved reliability, cleanliness, and pumping speed. They are effectively a small jet engine, driven by a motor. Some modern systems use a magnetically levitated turbine, limiting friction and particle generation from the bearings. Turbos can achieve pressures of less than 5×10^{-9} torr. An additional type of pump is the *titanium sputter pump*. Titanium is a very reactive metal, so during a sputter operation it reacts with many gas molecules, depositing the gas on the side of the chamber. In any sputterer with a titanium target, a short titanium sputter can dramatically decrease the vacuum pressure.

Vacuum system design is a complex subject because the way a chamber and its pumping lines are laid out can dramatically affect the way a system pumps. The foreline (the pump line leaving the large process chamber) and all its associated components (valves, traps, gages) are critical design parameters that can affect ultimate vacuum, cleanliness, pumping speed, reliability, and cost. All system dimensions depend on the pumps used and the desired operating characteristics of the final system.

5.2. Thin Films

To deposit a film of metal, oxide, or semiconductor onto a surface, the material must be made into a gaseous or *plasma* state. This can be done by heating the metal above its boiling point or by knocking atoms off of a surface using an ionized gas. Most *physical vapor deposition* (PVD) systems are designed to deposit metals. A second way to deposit a film is known as *chemical vapor deposition* (CVD). CVD is done by flowing a reactive gas across the surface, and supplying thermal or plasma energy to break down the gas, thus reducing it to a desired film material. Some metals, such as tungsten, can be deposited by CVD, but the most common films deposited by CVD are silicon oxides, silicon nitrides, polysilicon, and amorphous silicon.

Electroplating deposits metal films by reducing ions of a salt in a solution. *Electrophoresis* and *dielectrophoresis* use an electric field to pull polar or non-polar molecules out of a solution and make them stick to a surface. These techniques are not dealt with here.

5.3. Physical Vapor Deposition

Evaporators can be very simple systems, as shown schematically in **Fig. 1.9**. All that is needed is a chamber with a good vacuum, at least 10^{-6} torr, and a method to melt a metal. The two major types of evaporators are thermal and electron beam. In a thermal evaporator, a small quantity of the film forming metal is put into a boat, typically a refractory metal such as tungsten or molybdenum. A very large current is run through the boat, heating it up. The metal melts and evaporates since the chamber is free of gas. The better the vacuum, the purer the metal film deposited, so it is advantageous to pump for a long time before beginning the evaporation.

The second type of commonly found evaporator is the *electron-beam evaporator*. It uses a refractory metal filament to boil off electrons, which are then directed by a magnetic field into the boat containing the source metal. E-beam evaporators are more complex than thermal evaporators, but almost any metal can be evaporated, regardless of melting temperature. Commonly, carbon crucibles are used to hold the metal charge. E-beam systems may have large, moving substrate holders and are used in high-volume production environments.

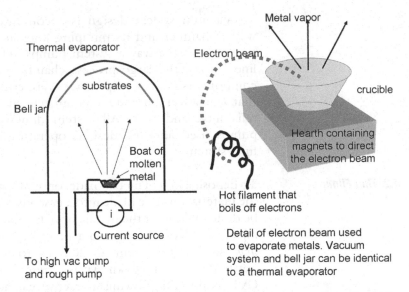

Fig. 1.9. Evaporator schematics. On the left is a simple thermal evaporator; on the right is a drawing of the hearth of an electron beam gun, which replaces the resistively heated source in the thermal evaporator. E-beams can even melt refractory metals. Carbon crucibles or refractory metal crucibles are often used to hold the source metal.

The only variables to control are the current through the heating element, the distance from the source to the substrate, and the pressure. One disadvantage of evaporators is that they always require that the substrate be above the source, thus fixturing is required to hold the substrate. Evaporation is a line of site technique, so sidewall coverage can be poor. Evaporators can have very high deposition rates and can deposit very pure and very thick metals. Heating of the substrates is rarely a problem, unless the source–substrate distance is too close.

Another way to deposit films is using a *sputterer*, such as that shown in **Fig. 1.10**. These systems are more complex than evaporators, but allow better control over the film properties. A metal target is attached to a "gun," usually a magnetron type. The magnets in the gun increase the plasma density near the target surface, increasing the bombardment of the target with positive gas ions, increasing the sputter rate. Targets are typically discs 2 or 3 inches in diameter and can be fairly inexpensive for common metals such as aluminum and titanium. For rare metals such as gold, sputterers are much more efficient in the use of metals than evaporators are. (Note that iron, nickel, gadolinium, and cobalt – the ferromagnetic metals – may be difficult to sputter using magnetrons.)

The atmosphere is evacuated and an inert gas at low pressure is allowed to backfill the chamber. The most common sputter gas is argon. A dc or RF plasma is generated between the gun and the substrate. Some machines sputter up, requiring fixturing for the

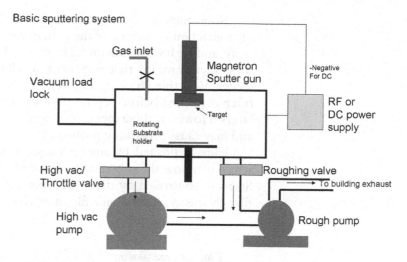

Fig. 1.10. A magnetron sputtering system. A vacuum load lock keeps the inside of the chamber at high vacuum while the substrates are being loaded. This improves vacuum quality and speeds up processing.

substrates, some sputter down. The plasma is often struck at a high pressure, 15–20 mt. After stabilizing, the pressure can be dropped to 2–3 mt for the actual deposition operation, since sputter rates go up as the pressure is dropped. Ions in the plasma knock atoms off of the sputter target by momentum transfer, creating a "gas" or "cloud" of atoms. These atoms then condense out on all surfaces inside the sputter chamber. Most machines also rotate the substrate during deposition so that the material coats the side walls of microstructures, thus sputtering has better step coverage than evaporation does. Sputtering has many more variables to control than evaporation, such as pressure, gas type, substrate temperature, dc bias, power, and source–substrate distance, which allow for finer control of film properties. Metals and conductors are usually sputtered with a dc power supply, from 50 to 500 watts. Dielectric materials such as oxides can be sputtered using a radio frequency (RF) power supply. Many materials can be sputtered that cannot be evaporated. Compound materials can be sputtered, but the composition and crystal structure of the resulting film may vary from that of the original target material. By introducing a reactive gas, such as nitrogen, materials such as titanium nitride can be deposited from a pure titanium target. Another way to make compounds in a sputterer is to run two targets simultaneously. To adjust the stress in the deposited film, some sputterers have heating capabilities in the substrate holder. Another advantage of sputterers is that the substrate can be cleaned by sputtering away some of the substrate material before depositing the desired film.

Noble metals such as gold and platinum are commonly used in micromachining because of their chemical inertness, high conductivity, and high work function. However, the chemical inertness of these metals means that they do not adhere well to many substrates. Very thin chrome, tungsten or titanium films (50 nm) are often deposited before depositing these metals to act as adhesion layers. However, these metals are more reactive than noble metals and may cause processing problems.

Films deposited by any technique may have stress in them, which can bow the substrate or cause films to crack (**Fig. 1.11**). Stress is controlled by deposition parameters, temperature, and film chemistry. Annealing a film may change its stress level.

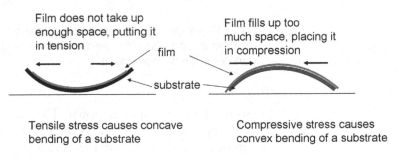

Stresses in deposited thin films and how they cause wafers to bow, relative to a flat plane- highly exaggerated

Fig. 1.11. Stresses that can develop in deposited films.

5.4. Chemical Vapor Deposition

The advantage of oxides, nitrides and polysilicon films for use in microdevices is that they are not reactive, are easy to deposit, and are widely available. The disadvantage of these films is that they generally require high temperatures, vacuums, and expensive equipment to deposit. Many chemical vapor deposition (CVD) processes rely on SiH_4, silane, as the source of silicon. This gas is pyrophoric, burning on contact with air. Some modern equipment uses this dangerous gas diluted in a carrier gas such as nitrogen or argon.

CVD is subdivided into many different technologies. Atmospheric pressure (APCVD), is good for oxides and requires ~350°C heat to crack the feed molecules. Low pressure (LPCVD) can deposit silicon oxide, silicon nitride, silicon, and tungsten, as well as silicides, and requires ~550°C heat to provide reaction energy, but provides high quality films. Plasma enhanced (PECVD) uses a plasma to provide the energy to crack the feed gas and can be done at temperatures as low as 100°C, but film quality may be poor at lower temperatures. It is best at depositing oxides and nitrides. Metal organic (MOCVD) is used for depositing compound semiconductor films epitaxially on crystalline substrates with an approximate lattice match to the film being deposited. CVD reactors need frequent manual cleaning to minimize particle deposition on substrates.

Plasma deposition equipment is very similar to plasma etching equipment, and both processes can be done in the same chamber, if the substrate is sufficiently heated and the right gases are plumbed in. Machinery dedicated to one or the other tends to work better.

In silicon processing, the best dielectric thin films are oxides created by *growing* a film on a clean silicon wafer at temperatures over 1,000°C by introducing oxygen or water. Dry oxygen gives the best quality films, but "wet" oxides grown with water can be made thicker because of the higher diffusion of water through the existing oxide. These films are often used as gate dielectrics and are not often needed in micromachining. These thermal oxides grow on the back of the wafer as well as the front.

As with all processes, you may be limited in what films you can deposit by the equipment available and substrate temperature limitations. Keep in mind thermal expansion issues as substrates and films heat and cool.

5.5. Electroplating

Electroplating can be thought of as the opposite of wet etching. If a *continuous* conducting film is placed in a bath containing ions of a metallic salt, metal ions can be reduced to metals on the conducting film as shown schematically in **Fig. 1.12**. It is usually done in

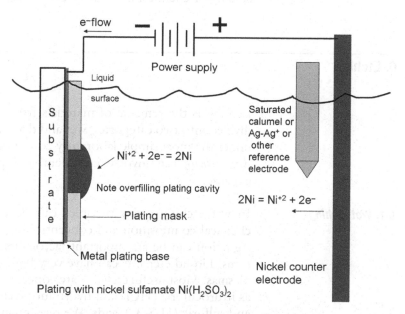

Fig. 1.12. A simple electroplating bath. Due to poor film uniformity, the geometries must be overplated. Final thickness is determined by a process that grinds off the top of the wafer until the correct thickness is achieved, known as chemical mechanical polishing (CMP).

aqueous media, but more reactive metals can be electroplated from organic solvents. The metals that can be deposited from any solvent must have a reduction potential less than that of the solvent. In other words, the water will break down into oxygen at the anode and hydrogen at the cathode before the metal ions will be reduced to metal at the cathode. Aqueous electroplating is usually limited to iron, nickel, copper, chrome, silver, and gold. It can be very cheap to do, requiring little more than a beaker, electrodes, solution, and power supply. Electroplating can deposit into very deep trenches, unlike sputtering and evaporation. Control of metallurgical properties is difficult because the plater has little control over crystal structure. Plating solutions contain many additives to control film properties. Very thick films can be grown this way, but uniformity is very poor, controlled by the non-uniform current density.

The IC industry has replaced sputtered aluminum with electroplated copper in what is known as the Damascene process, named after the famed Damascus swords with their swirled patterns in the metal. Copper conducts better than aluminum, and will not suffer electromigration (breaking of lines in high current applications due to atomic drift) but cannot be plasma etched. Holes are etched into an oxide down to a plating base and copper is plated up into the holes. The plating base is often a silicide, used to prevent copper from diffusing into silicon. After plating, the wafer is flattened, or planarized, using a process called lapping or chemical mechanical polishing, CMP.

6. Etching

Etching is the removal of materials from a substrate and can be divided into two categories, wet and dry. *Wet etching* can be done under the most simple laboratory conditions in a beaker, whereas *dry etching* can involve considerable capital and maintenance expense.

6.1. Wet Etching

In wet etching a few variables can be controlled, such as the chemical composition and concentration, as well as temperature. Agitation can be an important factor in some wet etching situations. Liquid etchants can have very high etch rates compared to plasmas. Most useful etchants are concentrated mineral acids, such as hydrochloric (HCl) and hydrofluoric (HF) and nitric (HNO_3) and sulfuric (H_2SO_4) acids. We recommend that these chemicals be purchased in semiconductor, electronic, or clean room grade. Etchants for some metals such as chrome can be purchased ready to use, sometimes etchants must be mixed in the laboratory. An excellent reference book is the CRC Handbook of Metal Etchants (1), which contains recipes for etching semiconductors as well as metals. Remember the AAA safety rule of *a*lways *a*dding *a*cid to water: never add water to an acid because the exothermic reaction can splash hot acids on you. Also, be very careful when mixing acids and organic solvents, as sometimes these mixtures can be explosive. Always dispose of acids and solvents in an environmentally acceptable manner. One driving force in industry to convert from wet etching to dry etching is the elimination of liquid wastes from wet etching operations, since environmentally sound acid disposal can be very expensive. Some factories have on-site acid recycling facilities. However, dry etching uses materials that are often discharged directly into the atmosphere, which can contribute to atmospheric pollution and potentially ozone depletion. With proper abatement equipment, such as burn boxes and scrubbers, plasma etching can easily meet environmental regulations, albeit at a rather high cost.

Hydrofluoric acid, HF, deserves special mention for safety. Although not one of the strongest acids, it is commonly used in device processing because it etches oxides and many metals. Sold at 49% concentration, it is almost always diluted prior to use as an etchant. HF does not immediately cause skin burns and it has the appearance of water. However, HF soaks through the skin and attacks bone tissue. It may leave a red rash on the skin, but little sign that it is doing damage deep within the body. It also reportedly attacks eye tissue, so when working with HF extra special care must be taken to keep this dangerous acid away from human contact. Never place HF into a glass container because it will attack the glass. HF can be safely used in Teflon or other fluoropolymer labware, polypropylene, or polyethylene. Since the cost of fluoropolymer labware is so high, we have had excellent results using inexpensive polyethylene plastic food storage containers with snap fit lids for storing and using acids and solvents in the clean room. These containers can be purchased at grocery stores or department stores. Although we have never specifically tested such cheap containers for contamination issues, we have never noted any problems caused by them.

Perchloric acid ($HClO_4$) finds some use in micromachining, and has been used to treat carbon nanotubes. It must be handled in a special hood, as its crystals are explosive.

The water present in HF oxidizes metals to their oxide, then the F^- ion reacts with the oxide to form a soluble byproduct. Sometimes HNO_3 (nitric acid) is added to acid etchants to oxidize metals, such as in aqua regia (royal water), a mixture of hydrochloric and nitric acids used to etch noble metals. HF/HNO_3 mixtures will attack silicon. Pure HF etches silicon oxides and is sometimes sold in a buffered mixture with NH_4F and surfactants in a 2:13 ratio known as buffered oxide etch (BOE) or buffered hydrofluoric (BHF). If you mix your own acids, be aware of the concentration. HF, for instance, can be supplied in many different concentrations in water, so take the diluent into consideration when mixing. Whenever any two acids are mixed, the reaction may be exothermic, so carefully consider the type of container used in mixing. Always wear heavy acid resistant gloves, eye protection and an acid-resistant lab apron and work in a fume hood.

In all etching, *selectivity* is an important parameter, the difference in etch rates between the material that is desired to be etched and other materials present that should not be etched, such as the etch mask and underlying layers. In wet etching, selectivity between the etch mask, usually photoresist, and the substrate can be almost infinite. In plasma etching, most materials present in the plasma will be etched to some degree. Wet chemical etching is usually an isotropic technique, with the material being etched horizontally as well as vertically at the same etch rate. This horizontal etching is responsible for much of the dimension change discussed above. The thicker the layer being etched, the larger the dimensional change (**Fig. 1.13**). Plasma etching can be controlled from anisotropic to isotropic etching, depending on the way the plasma is set up.

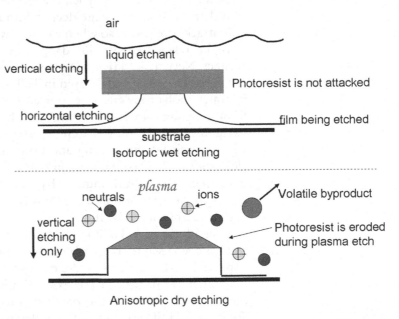

Fig. 1.13. Wet and dry etching. Wet etching is isotropic for most materials, etching in all directions at the same rate. Dry etching can be anisotropic, with the vertical etch rate many times higher than the horizontal etch rate. The etch mask is usually eroded during plasma etching.

In silicon and other crystalline materials, many liquid etchants have been developed that are highly specific to certain crystal planes or to the dopant level in the silicon. The classic is the warm KOH etch (**Fig. 1.14**). When used with (100) silicon, the (111) planes are etched very slowly. Details of KOH etching and some decorative etches can be found elsewhere (2, 3).

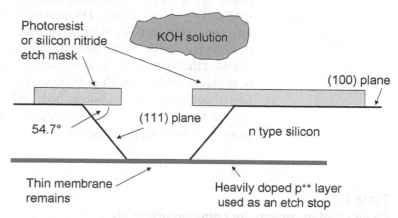

Anisotropic etching of crystal silicon to form a thin membrane

Fig. 1.14. Directional etching of silicon using KOH, (potassium hydroxide). KOH preferentially etches the (100) plane of silicon relative to the (111) plane. KOH also etches n (phosphorous) doped silicon much faster than p (boron) doped silicon, so a thin membrane can be made by etching through the back of the wafer to the p++ (heavily doped p-type) silicon, which is called an etch stop.

6.2. Metal Etchants

The metals most often encountered in device processing are aluminum, gold, platinum, and copper. Titanium, chrome, tungsten, and TiW (10%Ti, 90%W) are often used as adhesion metals underneath a noble metal. Nickel is used because it is easy to electroplate, but hard to sputter because it is a ferromagnet. Aluminum (usually as an aluminum ∼1% silicon alloy) is used in integrated circuits as a conductor because it is easy to deposit, has low resistance, does not tend to diffuse into silicon, and can be plasma etched. Copper was not traditionally used, as it tends to diffuse easily in silicon, putting states in the band gap. Copper cannot be plasma etched because there is no volatile copper compound. Copper has come into use recently using the Damascene process, which avoids etching (see above). Tin doped indium oxide (indium tin oxide (ITO)) 90% In_2O_3 + 10% SnO_2 is generally sputtered and is used as a transparent electrode in electroluminescent devices. Cobalt silicide is listed because it and other silicides are often put down under copper as a barrier to diffusion. Other metals are used, but they tend to be specific to an application.

An acid is only an acid in the presence of water (HCl is a gas when pure, but forms hydrochloric acid when water is added). The recipes that follow are generally based on the most concentrated chemicals available. For HF, this is 49%, H_2O_2 (hydrogen peroxide) 30%, HCl 38%, HNO_3 85%, H_3PO_4 (phosphoric acid) 85%, H_2SO_4 98%, NH_4OH (ammonium hydroxide) 28.5%, CH_3COOH (acetic acid) 98%. NaOH (sodium hydroxide) and KOH (potassium hydroxide) are sold as solids. They will attract moisture from the air, so keep the containers tightly closed. Recipes are specified by volume:volume ratios. These recipes are starting points. Individual recipes may be modified for specific effects. Dilution typically slows etch rate. Some mixes require water to work. Etch rates increase with temperature, but be very very careful when boiling an acid. As a general rule, one does not want the fastest etch rate available because films are rarely uniform in thickness, and very high etch rates tend to overetch one part of a film while another part is underetched. Etch rates can vary depending on how the film was deposited, purity, and heat treatments. A list of treatments is shown in **Table 1.1**.

Table 1.1
Basic etching recipes for common materials

Material	Etchants
Metals	
1. Gold	Aqua regia, tri-iodide
2. Platinum	aqua regia, chloroplatinic acid/lead acetate to oxidize platinum
3. Nickel	1:1 Nitric/acetic acid, HF/nitric mixes, CAN1
4. Copper	nitric/sodium chlorite, nitric/hydrochloric acids etch copper in various concentrations, as do $FeCl_3$ solutions.
5. Chromium-	Conc. HCl, commercial etchants, CAN2
6. Titanium	HF in any concentration, 20% H_3PO_4, 25% formic acid, 20% sulfuric acid
7. Tungsten	H_2O_2, 1:1 H_2O_2:HF, conc. H_2SO_4, 1:4 HNO_3:HF
8. TiW	H_2O_2, aqua regia, 1:2 NH_4OH:H_2O_2
9. Aluminum	10% NaOH or KOH, HCl, PNA
10. Indium Tin Oxide (ITO)	Conc. H_2SO_4, piranha etch, conc. HCl, 1 M oxalic acid, 55% HI (hydroiodic acid); see note below
11. Cobalt silicide	See note below. Any of the concentrated mineral acids mixed with H_2O_2 should etch the material.
Non-metals	
12. Silicon	1:1 to 1:10 HF:HNO_3, KOH (selective), HNA

(continued)

Table 1.1 (continued)

Material	Etchants
13. Silicon oxides – SiO_2, SiO_x	Dilute HF, BOE
14. Silicon nitrides – Si_3N_4	H_3PO_4 + few % H_2O at 160°–180°C. HF also etches this nitride, but etch rates vary depending on how the film was formed.

1. Aqua Regia – 3 parts HCl, 1 part nitric acid. Mixture may be explosive.

2. Tri-iodide – 400 g KI, potassium iodide, 200 g I_2, iodine solid, 1,000 ml water

3. *A special note on platinum. It is a difficult metal to etch, due to its inherent inertness. A common use of platinum in biology is as an electrode, usually coated with platinum black, which increases the current available through the electrode. Platinum electrodes are typically oxidized electrolytically in a 3% solution of chloroplatinic acid, $H_2PtCl_6.6H_2O$. The addition of a small amount of lead, copper, or mercury salt increases the available current, for example, lead acetate at 0.06% in solution.

4. CAN1 – ceric ammonium nitrate $(NH_4)_2Ce(NO_3)_6$ 50 g, 10 ml HNO_3, 150 ml water. Note that the Handbook of Metal Etchants gives the formula for CAN as $2NH_4NO_3.Ce(NO_3)_3.4H_2O$, showing one less NO_3^- group than the material that can be purchased from standard chemical catalogs. The four H_2O groups attached aid in dissolution, but the material is not generally sold as a hydrate.

5. Nitric sodium chlorite – 375 ml HNO_3 + 150 g solid NaO_2Cl plus water to make 1 l.

6. Chromium etchants are usually purchased mixed from a vendor that specializes in these etches. They are based on ceric ammonium nitrate and are designed for minimal undercut and high selectivity. Chrome is used for photomasks and as an adhesion layer, hence the need for good selectivity to gold and platinum.- CAN2 – ceric ammonium nitrate 10 g, nitric acid 100 ml, 1,000 ml water.

7. BOE or BHF, buffered oxide etch or buffered HF – typically contains 13:2 NH4F:HF or a similar ratio.

8. HNA – Hydrofluoric nitric acetic is the classic silicon etch. Nitric oxidizes the silicon, HF etches the oxide, and acetic acid is a pH buffer. Etching of silicon is so common that the reaction is given here: $Si + HNO_3 + 6HF = H_2SiF_6 + HNO_2 + H_2O + H_2(gas)$. The ratios are HF 8%, nitric 75%, acetic acid 17%.

9. PNA – Phosphoric nitric acetic – 80 parts phosphoric, 5 parts nitric, 5 parts acetic, and 10 parts water. Commercially available as a mixed acid, the most common aluminum etchant.

10. KOH – 7–8 Molar at 80°C with stirring. 450 g KOH/liter of water. Many concentrations will work, but 6–8 M have the best uniformity.

11. # ITO is generally plasma etched, often in CH_4/H_2 and argon mixtures, generally thought to be primarily a physical etch, rather than chemical. ITO is almost always deposited on glass, so any etchant for the doped oxide will also etch the glass substrate. Piranha etch is the same as the piranha clean, 4:1 to 10:1 $H_2SO_4:H_2O_2$.

12. +Cobalt silicide and other metal silicides are now used in silicon device processing and are usually plasma etched due to the small geometries generally used. Cobalt silicide has the highest conductivity of the silicides. Due to lack of volatile cobalt compounds, plasma etching is difficult, although chlorine and other halogen plasmas have been used with a heated substrate, as $CoCl_2$ is volatile at 200°C. CF_4/O_2 plasmas are also reported to work. We would also expect HF/HNO_3 mixtures to etch the film, but have not tested this mixture.

6.3. Dry Etching

Plasma, or reactive ion etching (RIE), etching is a complex subject, and the results obtained depend on many factors, including gas type, pressure, power, dc bias, electrode spacing, substrate type, and chamber configuration, as well as the ability of the machine to control pressure and gas flow. To the engineer designing a process, the etch equipment available and the gases plumbed into it ultimately determine what kind of etching can be done.

As with all microdevice processing, repeatable results depend on having machinery that operates consistently. During plasma etching, the byproducts of the etch are removed in the flowing gas and ideally exit the system through the pump exhaust. However, many etch byproducts are not volatile and deposit inside the chamber, in the pump lines, and in the pump. Most laboratories specify routine plasma cleaning of chambers using an oxygen/CF_4 mixture. This cleaning can remove only some deposits, so it is prudent to open the chamber on occasion and check for deposits on the walls and electrode, which can seriously degrade the etcher performance. Removing these deposits by scrubbing and wiping can have a large effect on the characteristics of the plasma. Some etchers introduce gas through a showerhead-type electrode. The holes can become plugged and must be cleaned out occasionally.

Plasma etchers come in many different styles and configurations. Simple machines such as barrel etchers create a plasma by wrapping a large cylindrical chamber with a radio frequency (RF) coil, usually operating at a standard 13.56 MHz. Plasmas are generated inductively though the glass chamber. These machines have low plasma density and non-uniform plasmas and are useful for resist stripping and surface modifications. Most control pressure by adjusting the flow rate of gas into the chamber, and typically only have one or two gases plumbed in.

One advantage of the inductively coupled barrel etcher is that no electrodes are required inside the chamber, and thus electrode materials cannot participate in the reactions that happen. Although considered an older technology, inductively coupled plasmas have come back into use in modern etchers, where a plasma is generated in an ICP (inductively coupled plasma) head and driven into the wafer by an additional potential applied to the wafer holder below the ICP head (**Fig. 1.15**). Much modern plasma etch equipment operates in this manner.

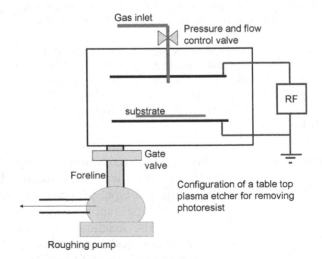

Fig. 1.15. A simple plasma etcher. Such systems often have only one pump and cannot reach pressures below 50 mt. These are best used for descum, stripping photoresist or activating a surface.

Another tabletop plasma reactor is a parallel plate etcher, which has a chamber large enough for a single wafer, and thus a higher density of plasma (**Fig. 1.15**). Most of these types of tabletop systems only have a single pump, and thus are limited to pressures above 50 mt.

Another configuration of modern plasma reactors uses an *electron cylclotron resonance* (ECR) chamber on top of the actual etch chamber to create a plasma, which is accelerated to the substrate by a powered substrate chuck, similar to the ICP reactor of **Fig. 1.16**.

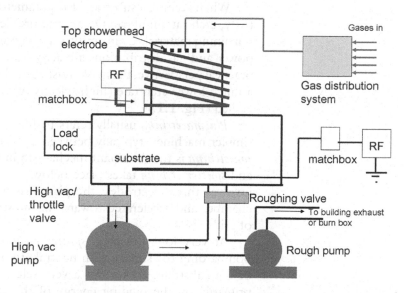

Fig. 1.16. A modern "deep reactive ion etcher" DRIE. The plasma is created using an RF coil above the wafer and driven into the wafer by an additional power supply on the wafer chuck. Such systems can etch trenches into wafers with vertical walls and high aspect ratios (up to 20:1).

Whenever a plasma is created with an RF power supply, a matching network must be in the system. This variable capacitor matches the feed from the power supply to the power reflected back to the power supply by the chamber. The reflected power should be less than 10% of the supply power. Many machines tune automatically, but some machines require manual tuning. Running a dummy wafer can get the tuning adjusted to an approximate setting before an important wafer is placed into the system. Do not operate a system that cannot be tuned properly.

Most ICP or ECR etchers have a load lock which loads wafers into the chamber without breaking vacuum. These robots only work with standard wafer sizes, so if you are etching a small or odd sized substrate, it may be necessary to put it on a larger wafer. Special pastes are made that contain a mixture of high vacuum grease and metal particles. These work well for gluing small pieces to a large substrate, since they provide both good thermal and electrical contact between the carrier wafer and the small substrate being etched.

6.4. Plasma Processes

During a plasma etch, three different processes can take place. Pressure is the primary determinate of which process dominates, with gas chemistry being a secondary variable. The relative contribution of each process depends on pressure, power, and temperature, as well as gas flow rate and type, and chamber configuration.

When etching a substrate, the parameters to control are selectivity, etch rate, uniformity, and the profile of the sidewall of the remaining material. Uniformity is determined by how uniform the power and gas distributions are across the wafer surface. From a practical perspective, the mask must either be much thicker or have a much lower etch rate (high selectivity) than the material being etched (**Fig. 1.13**).

Plasma etching usually refers to higher pressure processes in simpler machines, typically between 200 and 2,000 mt. *Reactive ion etching* is the dominant mechanism in the 50–100 mt range, and *sputter etching* takes place below ~50 mt. *Ion beam etching* involves three electrodes: the plasma is generated in a separate chamber and accelerated toward the substrate by a grid or series of grids.

At very low pressures, *sputter etching* removes material by bombarding the surface with neutral atoms, knocking out atoms of the substrate. It is a not a very selective process, depending primarily on the binding energy of the atoms in the substrate. Higher electrode potentials increase sputter etching at the expense of selectivity and possibly damaging the surface. Sputter etching can be very directional or anisotropic. Atoms sputtered away leave the vicinity of the surface due to their high energy and condense out of the plasma on chamber walls and in the pumping system. Almost any material can be sputter etched, but may contaminate the etch chamber. A common problem is that sputtered atoms can re-deposit on the wafer creating micro masks. The result is known as grass, a very rough surface in etched-out areas.

Chemical etching is exactly what it says: ions and radicals react with surface atoms creating a new *volatile* compound which is then removed in the flowing gas. Chemical etching is limited by the fact that the reaction byproducts must be in a gaseous state so that they can be pumped away. If the reaction leaves a solid material, it cannot be chemically etched using a plasma. Chemical etching tends to be isotropic, dependent primarily on gas composition and how effectively the plasma is ionized. The primary purpose of the plasma is to create ions and radicals, thus chemical etching is a weak function of power, but the etch rate tends to increase with pressure. The ratio between neutral and ionized gas species in a glow discharge plasma is something like 10^4–10^6:1.

Ion assisted or *ion enhanced* etching is found during simultaneous reactive etching and etching enhanced by ion bombardment. The etching of the substrate is chemical in nature and the reaction rate is determined by bombardment of energetic ions. Thus lower pressures and higher powers tend to increase the etch rate. The etch rate can be enhanced by the application of a bias across the plasma, but such capability is found only on advanced machines.

Highly directional etching, in silicon known as Bosch etching, takes place when an etch-inhibiting polymer is deposited on the substrate (2, 3). In one step, a polymer is deposited uniformly across the surface; subsequently, the chemistry of the plasma is changed to a reactive mode which attacks the polymer and the substrate. Since the polymer deposits uniformly, but the etching is primarily vertical, deep trenches with nearly vertical walls are possible with this technique.

Many different gases can be used in plasma etching. Oxygen is the most common, and reacts well with all polymeric materials. CF_4, sometimes known as Freon 14, is also very common, as are C_2F_6, C_3F_8, C_4F_8, and SF_6. The commonality is that all these gases are a good source of fluorine. Fluorine radicals are one of the most reactive chemicals known, and the only thing that can effectively react with oxides. Chlorine and its compounds are also frequently found in plasma etch systems, since many chlorides, particularly aluminum trichloride ($AlCl_3$), are volatile. Chlorine and bromine chemistries are common in compound semiconductor etching. Chlorine and fluorine chemistries are not typically compatible in a single chamber. Chlorine etched metals need to be handled carefully, as residual chlorine may remain on the surface once the substrate is removed from the vacuum. The chlorine can react with atmospheric moisture creating HCl which will corrode the metal pattern and substrate. Rinsing in water can remove the HCl. Treating the substrate with an oxygen plasma before removal from the vacuum will also alleviate the chlorine problem.

In *ion beam* etching in an ion mill, a plasma is created in a chamber above the substrate and accelerated toward the substrate with a grid or series of grids. These machines can have high etch rates and etch difficult materials.

One way to avoid etching metals is to do a *liftoff process*. With careful resist processing, the third profile shown in **Fig. 1.3** can be obtained. By putting the photoresist pattern where the metal does *not* go, and then evaporating a film, a metal pattern is defined. Sputtering does not work as well in liftoff process because it is less directional than evaporation and will coat the sidewalls of the resist pattern. The resist is then dissolved away in acetone, floating off the metal on top of the resist, leaving a well-defined metal pattern. The edges of the metal pattern may be rough.

7. Contamination Control

An area where there is much misunderstanding and myth is in contamination control. Simply by doing a process in a clean room in no way assures contamination free processing. Workers who understand contamination control can do much better work in a non-clean room than poorly trained workers can do in the best clean rooms. Industrially, billions of dollars are spent to eliminate every source of contamination, and extreme practices must be utilized to obtain and maintain tools, wafer handling equipment, people, and rooms at levels of cleanliness acceptable for the process. In smaller labs, such practices become burdensome and expensive. The level of cleanliness depends on the processes being carried out. One must decide what is "clean enough." The 10% rule applies in thin films as well as in x-y plane geometries – a defect 10% of the thickness of a film may cause problems with it.

There are several types of contamination. The most common is *particles*. Particles are ubiquitous in the environment, and the smaller the size, the larger the number of particles. They come from bacteria, people, abraded surfaces, aerosols, and especially the process equipment itself. Imagine breaking a biscuit or cookie in half. There are two large pieces, a few small pieces, and thousands of tiny crumbs. The smallest and most numerous particles cause the most problems and are hardest to eliminate. Preventing particles from getting onto a surface is much better than trying to remove them later. Thermodynamics requires that a clean surface become dirty, so a great deal of effort in any clean room is involved in removing the entropy increase that comes with contamination.

The second type of contamination is *ionic*. This source of contamination is not a large problem in micromachining, but is a significant problem for the integrated circuit industry. Ionic contamination comes from people, processes, and chemicals. High-grade chemicals made for the semiconductor industry are purified for trace elements at the factory, are filtered, and shipped in specially engineered bottles. These come with many different monikers, but all have specification sheets that tell what levels of ionic and particulate contaminants are present. "Off the shelf" chemicals, even spectrophotometric grade, are not as pure as chemicals made specifically for microelectronic processing.

The third type of contamination is known as *non-volatile residue* (NVR). All solvents have residues in them, even various microelectronic grades. It is easy to see how much is present by evaporating a drop of any solvent on a clean silicon surface and observing it under intense light. A single fingerprint on the inside of a chemical bottle, wafer boat, tweezers, or process chamber can add NVR to wafer.

Before spending a lot of money and introducing complex machines and procedures to reduce contamination, do a Pareto analysis. This is simply for identifying the major source(s) of contaminants and eliminating them first. The 80/20 rule often applies here – 20% of the effort will remove 80% of the contaminants. In industrial processes, the machine tools are probably responsible for most of the contamination, but in a small lab the people and all the surfaces that touch wafers are more likely to be the major source of contamination.

Since most substrate materials are dielectrics, they can easily pick up static electricity charges of 20,000 V or higher. This highly charged surface attracts oppositely charged particles. The only way to effectively discharge a polymer or glass surface is to spray it with an ionizing air or nitrogen gun. Spraying a surface to remove particles with an air stream that is not ion controlled is likely to charge the surface, actually increasing the number of particles on the surface. The overspray from a blow gun may stir up particles on surfaces making them airborne so that they land on wafers.

7.1. Counting Particles

There are now many machines made to count particles on surfaces and in fluids. Airborne particle counters can be obtained for a few thousand dollars, and corrosive liquid counters can be obtained for somewhat more. Both machines are reliable and helpful in monitoring the room itself and the fluids that touch wafers. If you have access to an airborne particle counter, turn it on and watch the counts as you move near it inside a clean room.

Particle counters for wafer surfaces are much more expensive and complex, especially those designed for patterned surfaces. Universities rarely have these tools. An inexpensive alternative is to shine a very bright light with a short wavelength on a surface. An excellent one is made by Spectroline and is sold as a "BlakRay," although any very bright light in a darkened room will do to look at wafers and other surfaces. It may expose photoresist.

7.2. Wafer Handling

In small research clean rooms, tweezers are commonly used to handle substrates. Fingers, no matter how well gloved, will put oils and particles on the surface. Substrates should never be touched by fingers. Vacuum wands are the optimal wafer handling tool since they only touch the back of the wafer, but are only effective in production environments. Hundreds of varieties of tweezers are available, so select stainless steel tweezers or tweezers with plastic jaws that are appropriate for the size of substrates you are using. Clean tweezers frequently with acetone and alcohol because any dirt on the jaws will be immediately transferred to your wafers. Store tweezers in a clean container, and do not use tweezers for levers, screwdrivers, and such, as nicks and damage to tweezers will scratch substrates.

Before a product wafer is put into any machine for processing, one or several dummy wafers should be run in the process to make sure it is running in a repeatable fashion. The previous user may have left some kind of contamination in the machine that will affect your process. For example, if you are depositing a metal film, the chamber could be coated with a metal that is incompatible with your process. In a sputterer, material on the walls may be sputtered away, depositing an unwanted contaminant in your film. By coating the chamber with the metal film you desire during a dummy run, your wafers will not be affected by the previous user's run. This is also true of plasma etch chambers and CVD deposition systems. Always do a dummy run before committing your product so that the chamber conditions are the same. Plan on having several wafers scrapped while you develop the process.

7.3. Clean Rooms

Simply working in a clean room will not assure particle free surfaces since few particles on wafers actually come from the air in a clean room. Most particles come from process solutions, process chambers, people, and dirty surfaces in contact with wafers.

Clean rooms are measured by their "class." The older classification is measured by the number of particles greater than 0.5 μm per cubic foot of air. A class 100 room has, for example, less than 100 0.5 μm particles per cubic foot of air in a room that has no people in it and has had time to come to a steady state. The modern classification is based on a metric standard, so a class 100 clean room in the older system is now a class 3.5 (3.5 particles greater than 0.5 μm per liter of air). Some universities have class 10 clean rooms, but most are class 1,000 or class 10,000. Clean rooms are always measured with no people in them.

The idea of a clean room is that air flows in a laminar flow regime, sweeping any particles out of the air and away from wafers. People and objects in the laminar air flow stream cause turbulence, which picks up particles which can then deposit them on nearby surfaces.

Regardless of the classification of the clean room, the HEPA (high efficiency particulate air) filters used to purify the air are common to all types of clean rooms and clean benches. If an airborne particle counter is used within a few cm of a HEPA filter, frequently no particles will be measured. After the laminar flow air passes equipment and people it picks up particles. Better clean rooms are built by controlling the way the air flows out of the room. For example, class 100 rooms often have sidewall returns, where class 10 or better rooms have air that returns through holes in the floor. Obviously, the better the clean room the higher the initial cost and the higher the maintenance costs.

Housekeeping is critical for clean rooms. The floor should be mopped with special mops and cleaners and should be vacuumed regularly with a HEPA filtered vacuum cleaner. Remove clutter. Position tables and machine tools so that air can flow around them – never place anything next to a wall, with the possible exception of perforated tables. Air must flow around a solid object on all sides to optimize cleanliness. Particles build up in dead air spaces and where the air rolls due to turbulence. Regularly clean tabletops, surfaces and doorknobs with clean room wipers and isopropyl alcohol.

7.4. Human Behavior in Clean Rooms

How people behave in a university type clean room may be *the most* important factor in how clean the wafers are, regardless of how much effort is put into engineering clean rooms and processes. Simply by dressing in clean room garments does not assure that people will not shed particles. In fact, making people dress in special garments is primarily a barrier to people entering a clean room. Even though gowned in a smock or jumpsuit (bunny suit), head covering, face covering, gloves, and shoe covers, you are still a source of particulate contamination. Having less people in a clean room is the best way to reduce particle counts, not by going to more exotic and expensive garments. Garments should have static control fibers woven into the polyester fabric, and the garments should be designed to prevent air puffs coming out of the sleeves and neck openings. As people move about in a clean room, the laminar air passes around people and objects picking up particles, which can then be deposited on surfaces. It is important that your head and hands never get above clean wafers. Move slowly in clean rooms. Never store wafers at floor level, even in closed boxes. Never put your feet up on a chair or table.

Many kinds of gloves are available, but ones designed for use in clean rooms have no powder on them to assist in donning and are frequently pre-washed. Always wash your hands before entering a clean room, even if you wear gloves, since finger oils can soak through most glove materials. Every surface your fingers touch will gain particles and oily films, and these contaminants can be moved around by diffusion (concentration gradients), by process fluids, and simply by putting wafers in contact with a surface that has been touched. For example, never place fingers inside a wafer boat, rather always hold boats by the outside.

Face masks designed to prevent particles leaving the mouth should be worn at all times. Never have wafers in front of you as you speak because particles come out of your mouth. Smokers should rinse their mouths out and wait 30 min before entering a clean room. Avoid cosmetics and colognes because these can be sources of particles and volatile organic residues. However, some skin creams can help prevent particles of flaking skin from falling off. Humans are the major source of sodium in silicon processing clean rooms. Sodium gets into the gate oxides of transistors and degrades their operating characteristics.

7.5. Cleaning

It is much more difficult to clean plastic, metal, and other substrate materials that cannot withstand piranha cleaning. Boats and boxes, tweezers, and other wafer-holding equipment must be cleaned regularly. Laboratory soap and a good rinse in DI water can effectively remove most surface contamination on wafer handling equipment, followed by an air dry. Do not wipe dry with any kind of wiper, as this will add many particles to the plastic surface.

An excellent cleaning procedure for equipment surfaces such as process chamber is the following:

1. Clean with high grade acetone using clean room wipers
2. Clean with high grade isopropyl alcohol (IPA) using clean room wipers
3. Clean with DI water
4. Vacuum

Mildly polar acetone is an excellent, cheap, and safe solvent, but contains a lot of residue, and may pick up more from the air. The more polar IPA removes the residue left by the acetone. Highly polar DI water is the cleanest and purest substance available (if properly piped in sterilized lines) and is the final clean to remove the residue left by the IPA. Do not get solvents into the waste water stream. A vacuum cleaner removes large lint particles left by the wipers. Be aware that the plastic tip of the vacuum cleaner nozzle can be a source of contamination on a newly cleaned surface, or melt on hot surfaces. Acetone, IPA, and DI water, followed by an effective drying technique is also a very good cleaning procedure for wafers that cannot take the more aggressive acid cleaning.

7.6. Wipers

When cleaning a substrate, table, process chamber, wafer storage container or photomask, the selection of wipers or towels is critical. Standard cellulose-based wipers commonly used in chemical laboratories have no place in microdevice processing. Many specialized types of wipers are available from clean room supply companies. Wipers are classified roughly as non-woven, woven, and sponges. One inexpensive non-woven type of wiper is a polyester cellulose blend. They have fairly low particulate shedding, are relatively inexpensive, and work well for general lab cleaning. Since sponges have no fibers, they are the best type to leave very clean surfaces, but are the most expensive. Sponges are recommended for mask cleaning. Mechanical wiping with a sponge can remove very small particles from surfaces, but can leave residues. Wipe a surface with a clean damp sponge, fold it over to expose fresh sponge and wipe again. Dispose of used sponge wipers after use.

Silicon wafers can be delivered from the factory with no particles on them. This pristine surface is obtained using a rotating sponge with copious amounts of water and cleaners. The sponge does not actually touch the surface. Few labs have such equipment, so a manual version can be done using sponge wipers. Any wiper made with fibrous materials will leave fibers and potentially scratch the surface.

Many clean room products such as wipers and chemicals come double bagged, having been packaged in a clean room. Open the box in an anteroom, remove the first plastic bag immediately before putting the object in the clean room, and remove the final bag once the item is in the clean room. This prevents clean products from getting dirty as they move from the dirty world outside the clean room to the clean area inside.

7.7. Facilities

In any laboratory, power and fluids must be delivered to the various wet processing hoods, plasma etchers and metal deposition equipment. The quality of the various processes strongly depends on the quality of the piping that delivers DI water and process gases. Even the tiniest amount of impurities or the tiniest particle will affect an integrated circuit. Gases such as CF_4 are delivered in special high pressure cylinders, in purities of 99.999% or greater. To get pure gases from the tank to the process chamber, the semiconductor industry has developed electropolished stainless steel orbital welded piping for process gases. Another advantage of such exotic tubing is that it will not leak and is highly resistant to corrosion, important safety factors. Specially trained workers with orbital welding equipment must be hired to install this kind of piping. Where a joint in such piping is required, special threaded fittings, sometimes known as VCR, which use a special metal crush washer. Standard collared fittings, sometimes known as Swagelok[R], or Gyrlok[R], are not acceptable for process gas lines because they can potentially leak and they can add considerable contamination to a process fluid. However, this type of stainless steel fitting is recommended for fluids that do not touch wafers, such as house nitrogen. A point of use filters on any fluid line is always a good idea.

When purchasing equipment, the cost of installation is sometimes overlooked, but it can be a significant percentage of the cost of machinery. We have seen very good and expensive equipment that never worked well because of poor quality and underfunded installation. Plan out the installation and factor in the cost of facilities, safety equipment, and pollution abatement when purchasing equipment.

7.8. Water

Pure water is one of the key elements in creating integrated circuits. Because deionized water is made on site and is continuously cleaned, it is the purest solution that any wafer will touch during processing. Good water systems are installed with special plastic piping that is welded together with special fittings. Deionized water is circulated continuously at a turbulent flow rate throughout the entire system. If water is very pure, any molecule, ion, or particle on a substrate put into it will see a large concentration gradient and find a lower energy state in the flowing fluid than on the surface. However, creating pure water from impure feedstocks and delivering it to the point of use is no trivial matter. Resistivity is commonly used as a measure of purity, but it is only one of several important water specifications.

Many small labs use small deionizing water systems (such as is shown in **Fig. 1.17**) that hang on a wall in the lab. Most of these systems make NCCLS/CAP reagent grade "Type 1" water with the following specifications:

Schematic of deionized water system for semiconductor quality water

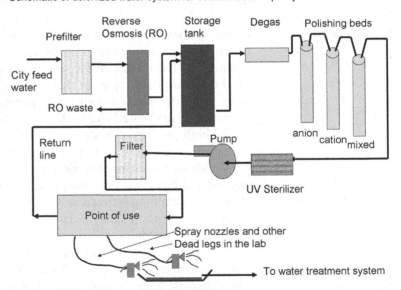

Fig. 1.17. A schematic of a continuous flow deionized water system sufficient to make semiconductor quality water. The prefilter removes large particulates from the feed water. The RO unit removes carbon and many other contaminants, but has a large pressure drop and creates a lot of waste water. The degas system removes oxygen from the water. The polishing beds remove ions and the UV sterilizer kills bacteria. After a final filter some of the water is used in the laboratory. The flow rate is sufficient to keep the water turbulent at all places in the piping, except the "dead legs." Dead legs are process tools and spray nozzles that do not have continuous flow and can be a source of bacteria. Unused water flows back to the storage tank and used water is sent to a water treatment system before being sent to the sewer.

Resistance – 10 MΩ

Dissolved SiO_2 – 0.05 ppm

Bacteria – 10 CFU (colony forming units)/ml.

Few laboratories do routine checks on such systems for bacteria. Wherever there is water that is not flowing, bacteria will grow. Unless sterilized routinely, water from these systems can be loaded with bacteria and dead bacteria pieces known as colloids. We do not recommend using water from this type of system. Even though the meter may read 10 MΩ or greater, there are many factors to consider in DI water cleanliness besides resistivity.

The semiconductor industry has developed extensive specs on water quality, driven by the lateral geometry size and the film thicknesses used for a given process. Such specs are also listed based on the number of bits on a memory chip, which is always the leading semiconductor technology. TOC is total organic carbon.

	2 µm/256 K memory	0.18 µm/1 G memory
Resistivity	>17 MΩ	18.2 MΩ
TOC ppb	<100 ppb	<1 ppb
Particles/l > 0.05 µm	–	<500
Particles/l > 0.10 µm	–	<100
Dissolved oxygen	–	<15 ppb
Bacteria CFU/ml	<100 ppb	<0.01 ppb
Dissolved silica	<10 ppb	<0.3 ppb
Cations	<1,000 ppt	<20 ppt
Anions	<1,000 ppt	<20 ppt

For most micromachine applications, the specs for a 256 K DRAM chip are adequate, although particulate may need to be reduced. Regardless of how pure and how carefully the water is sterilized at the purification equipment, *bacteria will grow in the piping system.* Microorganisms will grow in the rough surface inside the pipes, at joints in the pipes, and particularly in the dead legs where flow is only intermittent, such as in water spray guns. Filters are excellent places for bacteria to grow. As they die and become waterborne, they are particulates that end up on wafer surfaces. Consequently, all deionized water systems must be sterilized periodically, often with hydrogen peroxide.

7.9. Metrology

The accurate measurement of film thicknesses and lateral line widths requires specialized metrology equipment, most of it developed by the semiconductor industry. Besides the usual optical microscopes and scanning electron microscopes, the best tool for measuring micromachined devices is a profilometer. These tools gently place a stylus with a diamond tip or other hard material onto the surface of the device and drag it a specified distance across the surface. A readout gives a profile of the surface in the vertical axis and a distance readout for the in-plane features. To measure a film thickness, a step is required, which can be created by an etch or by placing a piece of tape on the device or test wafer while the film is being deposited. In metal deposition systems we often put a clean microscope slide with Kapton tape on it in the deposition chamber with the wafers to use as a metrology standard if there is no easy way to measure the film on a device wafer. The geometry size profilometers can image is limited by the tip diameter (**Fig. 1.18**).

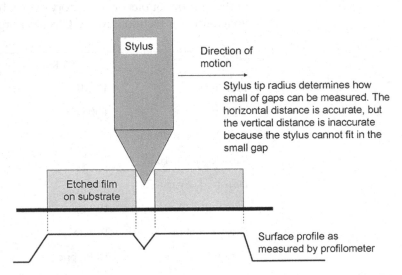

Fig. 1.18. Cartoon of a profilometer stylus that is too large to fit into a small gap, which causes an erroneous measurement of the hole depth. Horizontal distances are accurate, although machines made specifically for lateral dimensioning are probably more accurate than a profilometer. Profilometers exaggerate the *z* direction relative to the *x-y* plane. Even with profilometers it may be difficult to exactly determine the location of edges.

Many film thickness measurement tools use optical interference to measure film thicknesses, but they require knowledge of the index of refraction of the material. They work very well on oxides, nitrides, and polymers, but cannot work on metals. When calibrated correctly these tools can be very accurate. Other optical tools exist to measure line widths, but all have the limitation that the software must decide where the edge is, based on some kind of algorithm. All metrology equipment should be routinely checked against standards for accuracy and repeatability. All measurements have errors in it, and all films and linewidths vary across any substrate, so measure several point across your wafer to establish error.

8. The Traveler

Shown here is a sample run sheet, or traveler, for a very simple one layer process.

A model process run sheet
Wafer number_____ Date_____

1. Substrate
    a. Type_____
    b. Size_____
    c. Scribe numbers_____

2. Clean 1: piranha
    a. Time_____   Temperature_____   Acid mixture_____
    b. Spin rinse dry following clean_____
    c. Time_____ Date_____

3. Metal 1 deposition in sputterer No. 1
    a. Target number and type Titanium, gun 1
    b. Power_____Time_____Pressure_____
    c. Comments_____

4. Photoresist spin 1
    a. Resist type_____
    b. Spin speed_____ Acceleration rate_____
       Time_____   Adhesion   promoter used_____
    c. Resist thickness_____
    d. Soft bake time_____

5. Mask 1
    a. Exposure tool#_____ Intensity_____mw/cm^2
    b. Time_____

6. Develop
    a. Developer type_____
    b. Critical dimensions 1_____2_____3_____4_____5_____ Avg____ error____
    c. Time_____Temp._____

　　　d. Comments on
　　　resolution_____

7. Post exposure bake time_____

8. Metal 1 etch
　　a. Acid mixture_____
　　b. Etch time_____
　　c. Rinse time_____ Water resistivity_____
　　d. Drying step_____

9. Photoresist strip
　　a. Chemical used_____
　　b. Time_____

　　Process your wafers quickly. It is well known in industry that wafers which are processed faster tend to work better (higher die per wafer yield). In other words, reserve all the equipment you will need ahead of time and make sure that no PM (preventative maintenance) is scheduled or that the machine is down for repair. Get dummy wafers cleaned, warm solutions up to room temperature if necessary, clean up your work station, tweezers, boats, and boxes, and run dummy processes. Then begin to process your good wafers.

9. Odds and Ends

　　Most commercially available aluminum foils have peanut oil on them for processing. It is possible to find aluminum foil produced without oil, and it is recommended that only this type of foil be used in the wafer-processing lab. Such foil can be used to line spinners and vacuum systems and finds many uses in wafer processing. Kapton[R] polyimide tape and sheets is another very useful item to have in clean rooms because it can be used in vacuum systems. It has silicon-based glue that does not outgas. Other tapes are designed for use in cleanrooms and can be purchased from specialty suppliers. We recommend that only latex-laden clean room paper be used in clean rooms. These items and many other useful items can be obtained from clean room supply companies.

　　A step near the end of a process can affect earlier processing. Silicon processing has what is known as a thermal budget, accounting for all temperature steps in the process because heat effects are cumulative. Understand the impact of each process step on all the others from the beginning of the process.

Micromachine fabricators can take advantage of the large supply of used semiconductor equipment available on the market, although many suppliers provide new equipment quite suitable for micromachining. The cleanliness levels, throughput, uptime and the ability to control uniformity, and quality may not be as good as equipment designed for constant use in profit-oriented businesses. When equipment is purchased for a micromachining lab or small-scale factory, all these factors must be taken into account. Capital cost, however, is considerably less in smaller machines without automation, and such machines can be much more versatile. Quite a lot of this machinery is still in use in laboratories that were built 20 years ago. Cast-off machinery from production sources is often donated to university laboratories, but converting such machines to small-scale research may be more trouble than it is worth.

Often corporations can do higher quality processing than small research labs can, and for lower overall cost. For example, if you will need a lot of glass wafers covered with a metal film, it may be advantageous to order the wafers with a film already deposited on them, rather than to do it yourself. Ion implants can be done cheaper by contract facilities than can be done in house, so many labs have removed ion implanters.

A problem that may be encountered is the issue of cross contamination. Many labs are set up specifically to do silicon processing, which is *very* sensitive to ionic contaminants as well as particles. Lab owners may be reluctant to allow you to put non-silicon substrates with unusual films on them into process equipment. Unfortunately, we do not have an answer to this particular problem because a contaminated process tool can render hundreds of silicon electronic wafers inoperable. Determine what tools are available before finalizing your process run sheet.

10. Conclusions

The kinds of devices that can be fabricated with micromachining are only limited by your imagination, skill, the amount of money and time you have, and the machine tool resources available to you. However, microfabrication is notoriously difficult to get right because there are so many variables to control, and many important parameters are difficult to measure. Success is made much more likely by careful pattern design and layout, designing a process that can be built within the limitations of the toolset available, and by paying close attention to the details during processing. Even though much good science underlies all processing, in the real world processes are characterized empirically, so there is no substitute for experience with the various process tools. Consult with experts on process details and rely on manufacturers' specification sheets for details on photoresists, process chemicals, and adhesion promoters. Clean and characterize all equipment before committing an expensive and highly processed wafer to a process tool, and run plenty of dummy wafers to condition chambers and gather particles. Clean wafer containers and tweezers regularly, and keep your breath, body, and fingers away from devices. Move wafers through process lines quickly and use only fresh solutions. Plan on having many scrapped wafers while learning how to build your devices. Use idle time during pumpdowns and bakes to clean tables, equipment, and labware.

Acknowledgments

The author would like to thank Ken Danti, David Heredia, and Tracy Peterson for advice and editing. He would also like to acknowledge Sandia National Laboratories, a miltiprogram laboratory operated by Sandia Corporation, a Lockheed Martin Company, for the United States Department of Energy's National Nuclear Security Administration under contract DE-AC04-94AL85000.

References

1. P. Walker and W. Tarn (1990) *Handbook of Metal Etchants*, CRC Press, Boca Raton.
2. W.R. Runyan (1975) *Semiconductor Measurements and Instrumentation*, McGraw-Hill Book Company, New York.
3. M. Madou (2002) *Fundamentals of Microfabrication – The Science of Miniaturization*, Second Edition, CRC Press, Boca Raton.

Suggested Reading

1. Microlithography Fundamentals in Semiconductor Devices and Fabrication Technology, Saburo Nonogaki, Takumi Ueno and Toshio Ito, Marcel Dekker, New York, 1998.
2. M.L. Hitchman and K.F. Jensen, editors (1993) *Chemical Vapor Deposition, Principles and Applications*, Academic Press, Harcourt Brace Jovanovich, Publishers, Boston
3. V.M. Bright, editor (1999) *Selected Papers on Optical MEMS*, SPIE Milestone Series, Volume MS 153, SPIE Optical Engineering Press, Bellingham, Washington.
4. M. Gad-el-Hak, editor (2002) *The MEMS Handbook*, CRC Press, Boca Raton.
5. W.S. Trimmer, editor (1997) *Micromechanics and MEMS, Classic and Seminal Papers to 1990*, IEEE Press, Piscataway, New Jersey.
6. S.D. Senturia (2001) *Microsystem Design*, Kluwer Academic Publishers.

7. M. Ohring (2002) *Materials Science of Thin Films, Deposition and Structure*, Second Edition, Academic Press, San Diego.

8. J.L. Vossen and W. Kern, editors (1991) *Thin Film Processes II*, Academic Press, Harcourt Brace Jovanovich, Publishers, Boston.

9. S.K. Ghandhi (1994) *VLSI Fabrication Principles*, Second Edition, John Wiley and Sons, New York.

10. W. Kern, editor (1993) *Handbook of Semiconductor Wafer Cleaning Technology, Science, Technology, and Applications*, Noyes Publications, Park Ridge, New Jersey.

11. R.P. Donovan (2001) *Contamination-Free Manufacturing for Semiconductors and Other Precision Products*, Marcel Dekker, Inc, New York.

12. I.W. Rangelow (1996) *Deep Etching of Silicon*, Oficyna Wydawnicza Politechniki Wroclawskiej, Wroclaw.

13. S.K. Ghandhi (1994) *VLSI Fabrication Principles, Silicon and Gallium Arsenide*, Second Edition, John Wiley and Sons, Inc, New York.

Chapter 2

The Application of Microfluidics in Biology

David Holmes and Shady Gawad

Abstract

Recent advances in the bio- and nanotechnologies have led to the development of novel microsystems for bio-particle separation and analysis. Microsystems are already revolutionising the way we do science and have led to the development of a number of ultrasensitive bioanalytical devices capable of analysing complex biological samples. These devices have application in a number of diverse areas such as pollution monitoring, clinical diagnostics, drug discovery and biohazard detection. In this chapter we give an overview of the physical principles governing the behaviour of fluids and particles at the micron scale, which are relevant to the operation of microfluidic devices. We briefly discuss some of the fabrication technologies used in the production of microfluidic systems and then present a number of examples of devices and applications relevant to the biological and life sciences.

Key words: Lab-on-a-chip, dielectrophoresis, microfluidics, MEMS, microTAS, microfabrication, BioMEMS, biochip, microsystems, optical traps, PDMS, soft lithography.

1. Introduction

The field of microfluidics has been developing rapidly since its inception roughly a decade and a half ago. Growing out of the field of MEMS (micro-electro-mechanical systems) research, microfluidics has developed into a discipline in its own right, with a wide range of scientists and engineers now working in this area. The overall aim is the development of miniaturised biological or chemical analysis devices. The terms *"Lab-on-a-Chip," "uTAS"* (*Micro Total Analysis Systems*) and *"BioMEMS"* are synonymous and are often used when describing microfluidic devices with bio-chemical or medical applications.

M.P. Hughes, K.F. Hoettges (eds.), *Microengineering in Biotechnology,* Methods in Molecular Biology 583,
DOI 10.1007/978-1-60327-106-6_2, © Humana Press, a part of Springer Science+Business Media, LLC 2010

Microfluidic systems offer a number of advantages over more traditional macroscale laboratory techniques. The small size of microfluidic chips allows one to work with extremely small sample volumes (typically tens of nano litres as compared to the hundreds of microlitres required for microtitre plate based assays). Scaling-down affords saving in reagent cost and, more interestingly, is enabling the development of highly sensitive devices capable of isolating and analysing minute quantities of sample. Examples to date include microfluidic chips for single cell handling and analysis, single cell culturing chambers, microfluidic flow cytometers and cell sorters, and a variety of macromolecular detection and separation devices. Single-molecule measurement systems present another area where microfluidic and nanofluidic devices are offering great opportunities (1).

System integration is probably the most exciting aspect of microfluidic systems. The possibilities arising from the incorporation of multiple functions on a single chip, the size of a postage stamp, are astonishing, with sample purification, labelling, detection and separation all being performed as the sample is automatically guided from one region of the chip to another. A number of companies are currently producing microfluidic chips capable of performing routine assays and molecular separations. For instance, Caliper has developed a series of "*LabChips*" for DNA separation and cell analysis (2). These chips simply plug into a chip reader which then carries out automated analysis of the sample. Just as DNA microarrays have become ubiquitous in biomedical science, so too will microfluidic devices.

The number of journal papers that are published in this area has been increasing exponentially year-on-year, reflecting the growing importance of this field. A number of established journals have consistently contributed to the publication of microfluidics papers. These include *Analytical Chemistry, Analyst, Electrophoresis, Sensors and Actuators* (*A-Physical* and *B-Chemical*), *Langmuir, Journal of Chromatography A, Applied Physics Letters* and *Biophysical Journal.* Since its launch in 2002, the journal *Lab-on-a-chip* has focused solely on the publication of research in the field of microanalytical systems. There are also various conferences that are dedicated to the discussion of work in this area (e.g. "MicroTAS," "NanoTech-Montreux").

In this chapter we will present an overview of the physical principles governing the behaviour of fluids and particles at the micron scale. Theory that is central to the operation of microfluidic devices will then be discussed along with some of the fabrication technologies that are commonly used to produce microfluidic devices. Finally, a number of examples of microfluidic devices and applications relevant to biological and life sciences will be discussed.

2. How Fluids Behave at Small Scales

One aspect of reducing the dimensions in fluidic systems is that the fundamental physical properties change as the dimensions decrease. That is not to say that the underlying physics changes, but rather the relative influence of the forces involved. This means that scaling existing macroscale devices to the micron scale and expecting them to function well is often counterproductive. Various forces (e.g. electric, magnetic, optic, gravity) can be applied in the microsystem, allowing the manipulation of the fluid and particles suspended therein (e.g. cells, macromolecules). A number of excellent reviews giving detailed description of the physics of fluids on the micro- and nanoscale are available (3–5).

2.1. Laminar Flow

Scaling-down from the macroscopic fluid volumes, which we are familiar with in everyday life, to the microscopic dimensions found in microfluidic devices results in a reversal in the dominance of inertia and viscosity. In microfluidic channels, mass transport is dominated by viscous forces with inertial effects becoming negligible. As inertia is responsible for most of the instabilities and chaotic behaviour observed in fluids, its reduced influence within micron scale systems leads to what might be regarded as a simplified flow regime.

The Reynolds number is a useful concept and is defined as the ratio of the inertial properties of the fluid to the viscous properties and gives useful indication on the fluid flow behaviour. It can be calculated from the following expression

$$\mathrm{Re} = \frac{\rho \upsilon D}{\eta},$$
[1]

where ρ is the density of the fluid, υ is the velocity of the fluid, D is the hydraulic diameter of the particle and η is the fluid viscosity. The Reynolds number gives a measure of the characteristics of the fluid flow behaviour in the steady state and gives an indication of the relative smoothness of the flow (i.e. whether it is laminar or turbulent). For the case of a microfluidic channel, with at least one characteristic dimension in the micrometre range and with fluid flow velocities in the millimetre to centimetre per second (mm s^{-1}–; cm s^{-1}) range, flow is generally in the low Reynolds number regime (Re $\ll$ 1). At low values of Reynolds number the inertial effects which cause turbulence and secondary fluid flows are negligible and viscous effects dominate the dynamics of the fluid. This type of flow is termed laminar.

2.1.1. Pressure-Driven Flow

Hydrostatic pressure gradients, applied along the length of a microchannel, typically result in low Reynolds number flow. This combined with the non-slip condition imposed on the fluid at the channel walls results in Poiseuille-type flow, characterised by a parabolic velocity profile across the channel, with maximum flow velocity observed at the channel centre (*see* **Fig. 2.1(a)**). The parabolic velocity profile of the fluid has significant implications for particle transport within the channel. For instance, the position of a particle within the channel cross-section has a great influence on its velocity through the channel. A number of techniques have been developed to utilise this effect for particle separations (e.g. field flow fractionation techniques) (6, 7).

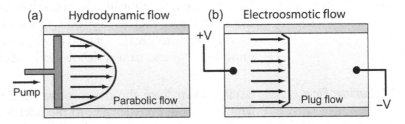

Fig. 2.1. Fluid velocity profiles resulting from (a) pressure-driven fluid flow and (b) electroosmotic driven fluid flow.

2.1.2. Electrokinetic Fluid Flow

Another method for creating flow in microchannels is electroosmotic pumping. An electrical potential is applied along the length of the flow channel. Charge on the walls of the channel results in the formation of an electrical double layer, with the build-up of oppositely charged ions near the walls. The ions move under the influence of the applied voltage (typically several hundred volts per centimetre) and drag the fluid with them, resulting in a bulk flow of fluid through the channel. The flow profile in this situation is markedly different from that of the pressure-driven system, with a uniform flow velocity profile being formed across the channel, as shown in **Fig. 2.1(b)**.

The flat velocity profile seen in electrokinetically driven fluid flow reduces the effect of particle position within the channel. However, sample dispersion due to band broadening can still be a concern, although it is not nearly as problematic as in pressure-driven systems. The implementation of electrically driven fluid flow on chip is straightforward, only requiring the dipping of electrodes into reservoirs at the ends of the microchannel. Furthermore, generation of the high voltages required is relatively trivial. However, the flow speed and direction are sensitive to the surface properties of the channel and sample composition.

2.2. Diffusion

In macroscale systems random eddies in the fluid continuously stretch and mix the fluid and as a result, transport and mixing time scales are greatly reduced. As the Reynolds number in microfluidic channels is typically below 1, no convective mixing takes place.

The only means by which solvents, solutes and suspended particles move other than in the direction of flow is by diffusion. For a spherical particle (e.g. molecule, cell) the diffusion coefficient, D (m^2 s^{-1}), is given by the following equation:

$$D = \frac{kT}{6\pi\eta r},$$ [2]

where k is Boltzmann's constant (1.38×10^{-23} J K^{-1}), T the temperature, η the viscosity of the fluid and r the hydrodynamic radius of the particle. If we look at the one-dimensional case, the distance x a particle will travel under the influence of diffusion in a time t is given by the following equation:

$$x = \sqrt{2Dt}.$$ [3]

The diffusion coefficient scales inversely with the size (the hydrodynamic radius) and shape of the particle. Small molecules have large diffusion coefficients and move further, on average, per unit time than larger molecules. Diffrences in diffusion coefficients have been used to separate different sized molecules and other larger particles over time (8).

2.2.1. T-sensor

Let us consider the case of two fluids flowing alongside each other in a microchannel. The laminar streams mix by diffusion alone, with any reactions taking place at the interface zone between the fluids. This interdiffusion zone spreads diffusively, increasing in size with time (and downstream distance). Solute molecules from each stream diffuse into the other and reactions can be monitored. The T-sensor shown in **Fig. 2.2(a)** takes advantage of this effect. T-sensors have been used to demonstrate a number of biochemical assays, these include the measurement of analyte diffusivities and reaction kinetics (9, 10), competitive immunoassays (11) and determination of analyte concentrations (12). Reactions taking place at the interface can be controlled accurately by varying the speed of the flows, allowing fine control of the reaction product (13).

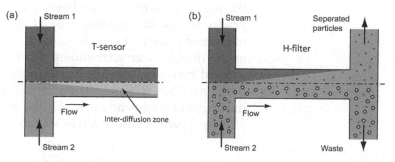

Fig. 2.2. Schematic of the operation of the (**a**) T-sensor and (**b**) H-filter.

The diffusion of molecules is affected by their size and therefore binding between antigens and antibodies can be monitored. By imaging the steady-state position of labelled components in a flowing stream, the concentration of very dilute analytes has been measured, requiring only a few microlitres of sample. High-speed assays have been demonstrated, with drugs, hormones and other small analytes being monitored using such devices, and detection limits below the 1–10 nM range have been demonstrated (14).

2.2.2. H-filter

Using the same principles as described above, the H-filter allows membrane-free filtering of particles based on size (i.e. diffusivity) (15, 16). A schematic of the H-filter is shown in **Fig. 2.2(b)**. Two streams enter the device from separate ports and flow alongside each other in a channel. Particle separation is achieved when the length of the channel or the flow rate of the sample is such that the particles of interest diffuse to fill the channel before exiting, whereas the other particles (those with lower diffusion coefficient) remain in their half of the channel and flow to waste. H-filters have been used to separate molecules based on differences in size (17) and to separate motile from non-motile sperm (18) as well as a number of other applications.

2.3. Electrokinetics

Electrokinetic methods are widely used in microfluidic devices. As a group of techniques, they all involve the movement of particles and liquids under the influence of applied electric fields (19). This movement arises from the interaction of the applied electric field with the particle and the fluid. The electric fields can be produced either externally (i.e. outside the channel) or locally by the use of microelectrodes patterned on the channel walls. A variety of forces and phenomena result, which can roughly be separated into those that act directly upon the particles and those that act upon the suspending fluid.

2.3.1. Electrophoresis

Electrophoresis is the motion of charged matter under the influence of an applied electric field. The direction of motion of a particle is towards the electrode of opposite charge; it is of no consequence whether the field is uniform or non-uniform. Due to the presence of the double layer, a charged particle when suspended in aqueous solution appears electroneutral. Despite this, movement of the particle still occurs due to the mobility of the ions in the double layer surrounding the particle. The velocity of a charged particle due to the electric field is, therefore, a function of the particle's size and charge, as well as the viscosity and conductivity of the suspending liquid. The electrophoretic mobility of a charged particle suspended in solution is given by the function

$$\mu_E = \frac{2\varepsilon\zeta}{3\eta} f(\kappa a),$$ [4]

where η is the viscosity of the liquid, ε the absolute permittivity of the liquid, ζ the particle's zeta potential, and $f(\kappa a)$ is a function of the ratio of the particle radius a to the double layer thickness κ^{-1}. The function $f(\kappa a)$ varies between 1 and 1.5 as the value of κa varies between 0 and ∞.

The main application of electrophoresis is in the separation of macromolecules such as DNA and proteins, with different macromolecules moving at different velocities under the influence of an applied electric field. As the technique will work with any charged particle, it is also applicable to the separation of cells, which typically have a net negative surface charge at physiological pH. Due to the simplicity of fabrication, a large number of capillary electrophoresis chips have been demonstrated in the literature (20), with most of these made by etching channels in glass substrates.

2.3.2. Electroosmosis

Electrohydrodynamic (EHD) forces on fluids arise as a result of the interaction of an applied electric field with the solution. Joule heating of the fluid causes the formation of gradients in conductivity and permittivity in the fluid. This variation in the electrical properties of the fluid and the applied electric field results in forces which act to move the fluid. Various "solid-state" pumps and mixers have been proposed based on the controlled application of EHD forces; these use carefully designed microelectrode arrays. See Morgan and Green (19) for a detailed description of EHD phenomena and applications in microsystems.

2.3.3. Dielectrophoresis

Dielectrophoresis (DEP) is the translational movement of neutral matter as a result of polarisation effects in non-uniform electric fields. DEP is different in nature from electrophoresis (EP) in that no net charge is required for movement to occur (19–23). Under the influence of a uniform DC electric field, a neutral particle will become polarised as a result of the electric field but will experience no net movement in the uniform field. If the particle is placed in a non-uniform DC electric field the situation is somewhat different. The charged particle will still experience a net translational force towards the electrode of opposite polarity. However, the neutral particle will now experience a net translational force, moving it towards or away from the regions of high electric field intensity, depending upon the relative polarisability of the particle and the suspending medium.

The magnitude of the force depends strongly on the particle's chemical and physical structure, shape and size, as well as on the frequency of the electric field. Consequently, fields of varying frequency can selectively manipulate certain particles with great sensitivity. This has allowed, for example, the separation of cells (24–27), bacteria (28, 29), viruses (30), DNA (31), proteins (32), carbon nanotubes (33) and other micro- and nanoscale objects (19).

In an AC field, the time-averaged DEP force $\langle F_{DEP} \rangle$ is given by (22)

$$\langle F_{DEP} \rangle = \pi \varepsilon_m R^3 \text{Re}\left[\tilde{f}_{CM}\right] \nabla |\mathbf{E}|^2 \qquad [5]$$

with the Clausius–Mossotti factor given by

$$\tilde{f}_{CM} = \frac{\tilde{\varepsilon}_p - \tilde{\varepsilon}_m}{\tilde{\varepsilon}_p + 2\tilde{\varepsilon}_m}, \qquad [6]$$

where R is the radius of the particle, $\mathbf{E}$ the electric field, Re[ ] represents the real part and $\tilde{\varepsilon}_p$ and $\tilde{\varepsilon}_m$ are the complex permittivities of the particle and the medium, respectively. In general, the complex permittivity of a material is given by

$$\tilde{\varepsilon} = \varepsilon - j\frac{\sigma}{\omega}, \qquad [7]$$

where ε is the permittivity, σ the conductivity, $j^2 = -1$ and ω the angular frequency.

According to equation [5], when the electric field is uniform, and therefore the gradient of the magnitude of the field is zero ($\nabla |\mathbf{E}|^2 = 0$), there is no DEP force. In the case of a non-uniform electric field ($\nabla |\mathbf{E}|^2 \neq 0$), the frequency dependence and the direction of the DEP force are governed by the real part of the Clausius–Mossotti factor. If the particle is more polarisable than the medium, ($\text{Re}\left[\tilde{f}_{CM}\right] > 0$), the particle is attracted to high-intensity electric field regions; this is termed positive dielectrophoresis (pDEP). Conversely if the particle is less polarisable than the medium, ($\text{Re}\left[\tilde{f}_{CM}\right] < 0$), the particle is repelled from high-intensity electric field regions and negative dielectrophoresis (nDEP) occurs.

2.3.4. Electrorotation (ROT)

The interaction of an electric field with a polarized particle creates an induced dipole moment. In a rotating electric field, a torque is exerted on the induced dipole which causes the particle to rotate. The time-averaged torque is given by (19, 22, 34)

$$\Gamma_{ROT} = -4\pi \varepsilon_m R^3 \text{Im}\left[\tilde{f}_{CM}\right] |\mathbf{E}|^2, \qquad [8]$$

where Im[ ] indicates the imaginary part.

Equation [8] shows that the frequency-dependent property of the ROT torque depends on the imaginary part of the Clausius–Mossotti factor. The particle will rotate with or against the electric field depending upon whether the imaginary part of the Clausius–Mossotti factor ($\text{Im}\left[\tilde{f}_{CM}\right]$) is negative or positive, respectively. The ROT torque is measured indirectly by analysis of the rotation rate (angular velocity) of the particle, which is given by (19, 35)

$$R_{ROT}(\omega) = -\frac{\varepsilon_m \text{Im}\left[\tilde{f}_{CM}\right] |\mathbf{E}|^2}{2\eta} K, \qquad [9]$$

where $R_{ROT}(\omega)$ is the rotation rate and K is a scaling factor. Again, owing to the viscous nature of the system, the angular velocity of the particle is proportional to the torque. ROT has been used widely to measure the dielectric properties of a variety of biological particles (35–41).

The frequency spectra of both the DEP force and ROT torque can provide significant information on the dielectric properties of biological particles in suspension. For further details on the theoretical relationship between DEP, ROT and dielectric spectroscopy, see references (42–44). A number of particle separation techniques have been developed using DEP (see below). It should be noted that the DEP force is of short range in its nature; its application is therefore well suited to microfluidic systems.

2.4. Surface to Volume Effects

The small cross-sectional area of microfluidic devices means that they have very high surface area to volume ratios. This means that microchannels are well suited for use in capillary electrophoresis (CE) applications, as they are very efficient at removing excess heat from the sample. A severe disadvantage of high surface to volume ratios is the problem of channel wall fouling. Cells and macromolecules diffuse to the walls of microdevices and adsorb onto the surfaces. This can reduce the efficiency of the electroosmotic pumping, resulting in complete channel blockage in extreme cases and generally interfering with the operation of the system. Optical and other sensors can also be disrupted due to build-up of such material.

3. Microfluidic Device Fabrication

A detailed description of the basic techniques used for fabricating microfluidic devices falls out with the scope of this article. Microfabrication technologies are covered extensively in the literature (45, 46) and within other chapters in this volume; we will therefore only briefly describe some of the more widely used of the current technologies.

3.1. Silicon- and Glass-Based Micromachining

Traditionally, silicon and glass have been the most commonly used materials for the production of microfluidic devices. The reason for this is mainly historical: micromachining of these materials had been widely explored in the context of MEMS research and it was therefore relatively straightforward to transfer this experience to the production of microfluidic devices. MEMS techniques allow complex systems to be fabricated out of silicon. However, silicon does have a number of drawbacks: it is expensive, it is difficult to machine (requiring access to well-equipped microfabrication facilities and highly trained personnel), it is optically opaque at wavelengths of interest (i.e. visible light) and it is generally not well suited to biological applications due to its surface characteristics and bulk electrical properties. Glass is commonly used as a material in microfluidic devices with biological applications; however, it suffers from some of the same drawbacks as silicon. Both these materials do, however, lend themselves to applications where strong solvents or high temperatures are required. Glass dominates in the area of on-chip capillary electrophoresis where the surface properties of the glass channels are important in determining the flow properties of such systems.

3.2. Polymer-Based Microfluidics

The use of polymers has grown rapidly over the last few years due to their low cost and the development of a number of reliable fabrication methodologies (47). The reduced cost of such devices allows for the possibility of making them disposable, a factor crucial for the uptake of microsystems technology in many medical, diagnostic and biotechnology applications where sterility is important. The area of polymer-based microfluidics can be broken down into four main technologies: polymer micromachining, soft lithography, micromoulding and in situ construction.

3.2.1. Polymer Micromachining

Polymer micromachining uses many of the same technologies as that of silicon-based fabrication and tends to require access to at least simple cleanroom facilities such as resist spinners, mask aligners and solvent development tanks. Thin layers of photoactive polymers are either spin coated or laminated onto a solid substrate, typically glass or polymer. These layers are photolithographically patterned, resulting in the selective cross-linking of the polymer. The channels are then revealed via solvent-based development (i.e. removal of uncross-linked polymer). The most commonly used materials are epoxy-based polymers such as SU-8 and several of the polyimides. Fabrication of complex three-dimensional channel geometries is possible. These materials when fully cross-linked are also largely resistant to chemical attack. Madou describes a number of polymer micromachining techniques (45).

3.2.2. Soft Lithography

The use of soft materials (polymers and gels) has a number of advantages over that of silicon and glass. The term "soft lithography" is used to describe a set of fabrication techniques distinct from what we have termed above as polymer micromachining. Soft lithography is based on the use of two techniques: moulding and embossing.

3.2.2.1. Polydimethyl siloxane and Moulding

The use of polydimethylsiloxane (PDMS) has revolutionised the field of microfluidics. This technology, originally introduced by the group of Whitesides (48), has largely been responsible for opening up the field of microfluidics to those without access to high-tech microfabrication facilities. PDMS is a silicon rubber and is the material of choice for many micromoulding applications. It is inexpensive, biocompatible and transparent, making it ideal for use in many biochemical applications. PDMS also has excellent sealing properties both to glass and also to itself, the latter allowing for the fabrication of complex multilayered fluidic structures (49). Typically, a mould (commonly known as a "master") is produced using one of the hard micromachining technologies described above (depending on the material, surface treatment of the master may be required to facilitate delamination of the cured PDMS from the master). The PDMS prepolymer (mixture of the PDMS linear polymer and a cross-linking agent) is mixed and then poured onto the mould and left to cure. Once the polymerisation is complete, the cured PDMS is peeled off the master and can be bonded to a solid substrate (e.g. glass) or another piece of cured PDMS to form the fluidic channel. Surface treatment of the PDMS is possible, allowing one to change its wetting properties (from hydrophobic to hydrophilic), as well as other surface functionalisation techniques (50, 51). A number of review articles by Whitesides and others are available (48–50).

3.2.2.2. Hot Embossing

Polycarbonate and PMMA are the most widely used polymers for embossing (45, 52, 53). A stamp (master) is brought into contact with the polymer surface. An even pressure is applied and the polymer substrate is heated above its glass transition temperature, the polymer flows taking up the profile of the master. The substrate is cooled and removed from the stamp. This technique can be used to produce micro- and nanoscale features and is described as micro- or nanoimprint lithography (54). As well as microfluidic channels, embossing has been used to fabricate a number of optical components such as microlenses, diffraction gratings and waveguides (55, 56).

3.2.3. In Situ Construction

The incorporation of hydrogels and other similar polymers into microfluidic channels is a recent development (67–60). The integration of such active polymers with other micromachined and soft materials allows for the implementation of environmentally (i.e. within the microfluidic channel) responsive functionalities that are difficult to realise otherwise. Work has shown the use of polymers that react by swelling in response to changes in the local environment of the fluidic channel. Hydrogels sensitive to pH, temperature, conductivity, light, glucose level, etc., have been demonstrated, and applications include their use in "smart" on-chip fluid flow control (57–61).

3.3. Other Methods

A number of other technologies have been explored within the literature. Microstereolithography is a technique allowing the fabrication of three-dimensional structures; it relies on the photopolymerisation of liquid polymers using a focused beam of UV light (62). Powder blasting has been used to fabricate channels with dimensions down to 50 µm; this technique is limited by the rather rough channel surfaces created, resulting in difficulties in observation (63).

A number of examples of channel-free microfluidic devices have been presented; these can generally be described as droplet-based microfluidics, where the sample volume is confined within droplets formed in two-phase systems (typically such experiments are carried out in water-in-oil or vice versa to reduce the effect of evaporation). Examples can be found in electrowetting technologies (64, 65) and droplet-based chemical reactors (66).

3.4. Biocompatibility

Surface chemistry is of great importance in microfluidic systems due to the extremely high surface area to volume ratio, with non-specific adsorption of sample to the channel walls resulting in channel blocking and disruption of the fluid flow (especially in electrokinetic-based flow devices). As described above, silicon does not generally lend itself to chemical modification and therefore is limited in its utility. Glass and various polymers have good properties and are widely used, with polymers generally being preferable due to their reduced cost over glass. PDMS has the highly desirable property of being gas permeable, thus making it suitable for cell culture applications. For a recent review on biocompatibility in microsystems the reader is referred to the literature (67).

4. Applications

We will now look at some examples of microdevices used in biology. The selection of devices represents a diverse range of applications, but is by no means exhaustive in its coverage.

4.1. Cellular Manipulation and Analysis

A number of on-chip cell detection and handling techniques have been demonstrated in the literature. Examples include cell culture devices, electroporation and lysis chips and microfabricated flow cytometers. Working with small numbers of cells also becomes possible and recently a number of single-cell devices have been developed.

4.1.1. Cell Culture Devices

Studies of drug effects, osmotic balance, cytogenic and immunologic responses and metabolism have all been carried out using microfabricated cell culture devices (68). The first human embryonic stem cell culture in a microfluidic channel was recently reported by Abhyankar et al. (69). Integrated chips with embedded fluidic channels for nutrient delivery were reported by Heuschkel et al. (70). A number of systems demonstrating the possibility of long-term cell culture in microfluidic networks have been reported, and the effects of nutrient levels and oxygenation (71). Chin et al. (72) demonstrated a massively parallel single cell array consisting of 10,000 microwells, all of which were exposed to the same media. Over 3,000 individual rat neural stem cells were cultured simultaneously. The cells were able to draw signalling factors released from neighbouring cells within the device and were also able to maintain paracrine signalling throughout experimentation. Single cells were shown to survive and proliferate, demonstrating the importance of intercellular signalling on cell viability. Microfluidic devices capable of long-term culture of multiple isolated cells, separated in individual wells, have obvious application in drug testing or cell growth analysis (73, 74).

Figure 2.3 shows a multiplexed cell culturing device as reported by Lee et al. (73). An addressable 8 × 8 array of 3-nL chambers was fabricated and used to observe the serum response of HeLa cells in 64 parallel cultures. The device is capable of producing concentration gradients along the length of the device and individual rows are separately addressable.

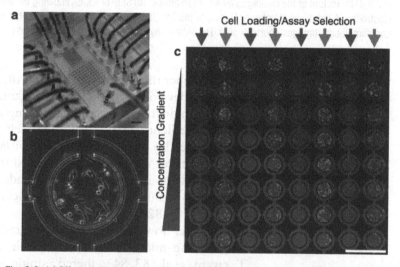

Fig. 2.3. (a) Micro cell culture device with a concentration gradient generator. (b) HeLa cells cultured in a single well of the array; the central growth area has a total volume of 3 nL. (c) Fluorescence image of the array after 5 days of culture (73). (Reproduced with permission from John Wiley & Sons, Inc.)

Using photo-thermal etching of agar, it is possible to modify the culture chamber geometry during culture. Studies have demonstrated that it is possible to construct dynamic 3D networks of cells, with no adverse effects on the cells under culture (75).

The inclusion of microelectrode arrays (MEA) in the base of cell culture chips allows for the electrical recording and stimulation of the cells under culture. **Figure 2.4** shows an example of such a device. Measurements of the localized electrical signals from neurons and cardiac cells have been performed with application to the study of drug effects on ion channels (76–78). The possibility of culturing cells on multiparametric integrated sensors on silicon was demonstrated by Brischwein et al. (79).

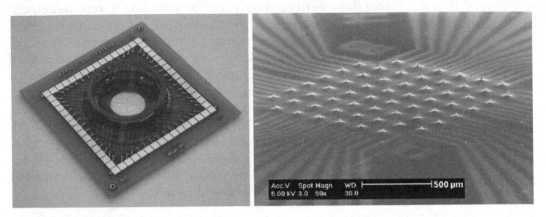

Fig. 2.4. (a) Picture of the packaged MEA culture device. (b) SEM of a MEA showing the 60 tip-shaped protruding platinum electrodes. A thin layer of polymer is used to insulate the electrode tracks so that only the tips are exposed for recording and stimulation. (Images courtesy of M. Heuschkel, Ayanda Biosystems, Switzerland.)

A number of systems for studying the effect of drugs on cells have been developed. Tools for cell docking and capture have been used in conjunction with controlled drug application and concentration gradients (80). Patterned self-assembled monolayers (SAMs) are also used to control cell surface topology and molecular structure. Whitesides' group used these techniques to help understand the interaction of man-made surfaces with cells and proteins (81) and for the study of cell conformation, attachment and function (82).

Laminar flow can easily be used to create spatially and temporally varying microenvironments within the cell culture device. Takayama et al. (83, 84) adhered a single bovine capillary endothelial (BCE) cell in a microfluidic channel and flowed Trypsin/EDTA locally over it and observed the detachment of the treated area of the cell from the channel floor. The simple microfluidic device consisted of three inlets from which the flows converged into a rectangular capillary as parallel laminar streams. Different fluorescent dyes were flowed over opposite sides of a live cell, staining two subpopulations of mitochondria for observation. The cell was locally treated with latrunculin A and the mitochondrial response to the cellular damage was observed.

4.1.2. Electroporation
and Cell Lysis

Electroporation exposes cells to electrical pulses of high intensity (10^6 V m^{-1}) and of short duration (10^{-3}–10^{-6} s), causing a temporary increase of the cell membrane permeability. This leads to ion leakage, escape of metabolites and increased uptake of drugs, molecular probes, DNA, etc. (85). Applications of electroporation include the introduction of plasmids or foreign DNA into living cells for transfection and the insertion of proteins into cell membranes. Generally performed in large vessels, this technique suffers from a relatively low efficiency due to that fact that all cells are exposed to the same electric fields. On-chip electroporation has been demonstrated (86, 87). The application of nanosecond pulsed electric fields (nsPEF), with high intensity and low energy, to single cells allows the targeting of intracellular membranes and has been shown to induce a number of cellular apoptotic or non-apoptotic functions without directly harming the plasma membrane (88).

Single-cell electroporation is easily achievable in microfluidic devices and permits the tuning of the applied electric pulse for the individual cell according to its size and other electrical parameters, giving more accurate control of the transmembrane voltage. The composition of the medium directly around the cell is of prime importance and can be closely controlled during the electroporation process and pore opening time. For instance, the influx of Ca^{2+} into the cell should be kept low. Reactive oxygen species (ROS) liberated by the cell during the electroporation process also impact on the cell survival rate. Microfluidic studies have demonstrated that ascorbate (a ROS scavenger) can reduce the cellular damage induced by ROS and increase cell survival rate up to 50% (89).

Cell dipping and washing using controlled fluid flow and nDEP was recently demonstrated on a chip (**Fig. 2.5**) (90). Cells flow into the main channel of the device from one of the inlet channels. The reagent of interest is flowed in through a second inlet and flows side by side with the cell sample in the channel. Electrodes on the walls of

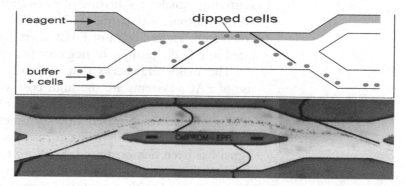

Fig. 2.5. (**a**) Schematic of cell flow dipping in a microchannel. (**b**) RBCs suspended in PBS are deviated laterally into an adjacent stream. After a short exposure to the other stream, the cells are brought back into the PBS stream. (Images courtesy of Nicolas Demierre and Urban Seger.)

the flow channel guide the cells across the channel into and out of the reagent stream. Flow ratios and speeds are set such that no dye contaminates the cell outlet and to permit nDEP cell manipulation. Varying the flow rate allows control of the incubation time of the cells in the reagent under study. Very rapid reaction and washing cycles are possible with such a device.

4.1.2.1. Cell Lysis

Cell lysis is a preliminary step in the analysis and separation of intracellular components such as DNA and proteins. Lytic agents such as detergents or pure water are used to rupture the cell membrane. A change in the tonicity of the suspension medium is typically used to lyse erythrocytes and produce ghosts (spherical emptied RBCs), which can then be resealed. On-chip electrolysis has been demonstrated and can be performed in a timely, selective and localized manner with reduced risk of damage or alteration of the cellular content before separation (91, 92). Single-cell enzymatic lysis and identification of β-galactosidase activity have been demonstrated (93).

4.1.3. Microflow Cytometry

For single-cell optical measurement and sorting, the fluorescence-activated cell sorter (FACS) offers a broad range of possibilities (94). Flow cytometry allows the simultaneous measurement of multi-colour fluorescence and light scattering from individual cells as they rapidly pass through one or more focused laser beams. Light absorption and scattering can be used to determine relative cell size, shape, density, granularity of a particle as well as stain uptake due to labelling with fluorescent probes. Modern benchtop flow cytometers allow high-speed analysis and cell sorting, with sort rates in excess of 10,000 cells per second possible with modern machines.

It has been recently pointed out that even for high-end versions of commercially available FACS instruments, forward scattering signals of a mixture of size-calibrated beads is strongly non-monotonic with particle volume (96). This issue seems to be completely ignored by most FACS users, who are generally interested in broadly positive or negative fluorescence decisions.

The major drawbacks that currently limit the widespread use of FACS systems are the high cost, complexity and size of the instruments. To this end, a number of groups have been developing microfabricated flow cytometers. On-chip optical detection of single cells using forward scatter or fluorescence signal has been demonstrated. Some of these systems have also integrated sorting functionality, although at much reduced rates compared with commercial instruments. Rare-event cell sorting in a microfluidic system was suggested as a safer alternative method for the enrichment of prenatal nucleated RBC from maternal blood (97).

Early results obtained using a microfluidic-based haematology analyzer using small-angle scattering information from hydrodynamically focused leukocytes showed possible differentiation of leukocytes (98). Schrum et al. presented a microchip flow cytometer that uses electrokinetically generated sheath flows as an alternative way to achieve particle focusing (**Fig. 2.6**) (95). A microfabricated fluorescence-activated cell sorter that can achieve rates up to 20 cells·s^{-1} was demonstrated by Fu et al., based on a fast forward–backward electrokinetic flow switching procedure in a T-shaped channel structure (99). A microfluidic device for fluorescence-activated cell sorting that incorporates integrated light sources, sensors and micro-optical components was discussed by Kruger et al. (100). Measurement of the intrinsic autofluorescence signal of single cells on chip showed measurable differences in the fluorescence levels between erythrocytes and granulocytes (101).

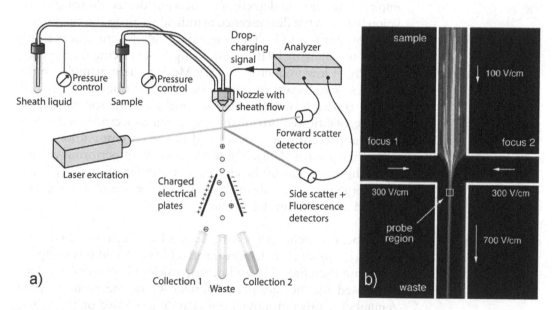

Fig. 2.6. (a) Schematic of a fluorescence-activated cell sorter. (b) Time-integrated CCD image of electrokinetically focused 1.88 μm labelled particles on microchip. The exposure time was 5 s with sample and focusing field strengths of 100 V cm^{-1} and 300 V cm^{-1}, respectively. Arrows depict the direction of fluid transport and their lengths are proportional to average fluid velocities in each channel. (Figure (b) reproduced with permission from Schrum et al. (95), © 1999 American Chemical Society.)

Optical tweezers using tightly focused laser beams provide a tool for the manipulation of single micron-sized particles (102). The technique is commonly used in microchips to capture beads, cells and DNA (103). Complete integration of such a system into a portable unit presents some issues (laser power source, movable optics, etc.). DEP techniques are often favoured for control of particle trajectories within the flow. Moreover, simultaneous manipulation of multiple objects using the laser tweezer technique is difficult. However, newly developed holographically generated optical traps have demonstrated the possibility of generating multiple optical traps in three dimensions (104).

The first μCoulter devices were presented by Larsen et al. (105) and Koch et al. (106, 107). These systems were microfabricated in silicon and use microfluidic sheath flows and filter structures. These devices allow the measurement of the electrical properties of individual particles as they flow through a microchannel. Reports of more sophisticated on-chip impedance measurement devices were published later by Fuller et al., who looked at granulocytes (108), Larsen et al. studied somatic cells using a simple silicon aperture (109) and Gawad et al. used a coplanar and facing electrode geometry to measure the properties of erythrocytes (110). More recently, Benazzi et al. investigated the properties of different species of marine algae; this work also implemented optical detection in the same device allowing correlation between the fluorescence of individual algae and their impedance properties (111). The original μCoulter technique is clearly limited to cell sizing by the use of a single measurement frequency, generally at low frequency or DC. Measurement at multiple frequencies is necessary to determine other attributes of the cell, similar to what is obtained in traditional dielectric spectroscopy.

Microfabricated broadband single-cell dielectric spectroscopy chips have recently been developed by the current authors (112) and other groups (108). These are capable of performing high-throughput analysis of the electrical properties of single cells and other organisms. Details of this work can be found in Chapter 7 (Gawad et al.) within this volume.

4.1.4. Cell Sorting

A number of on-chip cell-sorting devices have been described in the literature. Quake et al. demonstrated a PDMS-based FACS capable of sorting bacteria, DNA and other particles into two outlet channels based on the measured fluorescence of the particle (104). A number of other groups used similar devices based on the switching of fluid flows between outlet channels using electrokinetic flow.

Holmes et al. (113) demonstrated devices for high-speed analysis and sorting of individual cells and polymer beads based on their optical properties. The device used nDEP to focus the particles onto the central flow axis of the channel as they flowed through the device. They obtained results similar to that of commercial FACS, although at reduced particle throughput (100 s of particles per minute). An arrangement of electrodes at the channel junction is used to allow fast sorting of particles into one of two outlets, using nDEP. Fluorescently labelled beads demonstrate the system's performance. Similar work was carried out on samples of blood cells, bacteria and algae. A number of other DEP-based particle sorting devices have been reported (7, 27, 114–116).

4.2. Macromolecules

The area of micro total analysis systems that has seen the largest number of applications is that of molecular separation, in particular the separation and analysis of subcellular components such as proteins and DNA. On-chip separation techniques making use of electrophoresis include capillary electrochromatography (CEC) (117), which typically employs microbeads packed in the capillary channel. Isoelectric focusing (IEF) is a commonly used technique, whereby a molecular-specific isoelectric equilibrium point is reached in a pH gradient (118, 119). Dielectrophoresis is widely used to separate proteins, DNA, viruses and other bio-nanoparticles (19).

4.2.1. DNA

4.2.1.1. PCR

One of the most common methods for analysing DNA and proteins is capillary electrophoresis (CE). In the case of DNA it is generally preceded by an amplification technique: the polymerase chain reaction (120). **Figure 2.7** illustrates an example of an on-chip PCR device; a serpentine channel was fabricated to flow across three heater elements with temperatures of 95°C, 77°C and 60°C. The flow rate, length of the channel and serpentine geometry define the duration, rate and number of heating cycles that the sample undergoes. Microdevices that intergrate cell lysing, PCR and CE functionalities have been demonstrated (121).

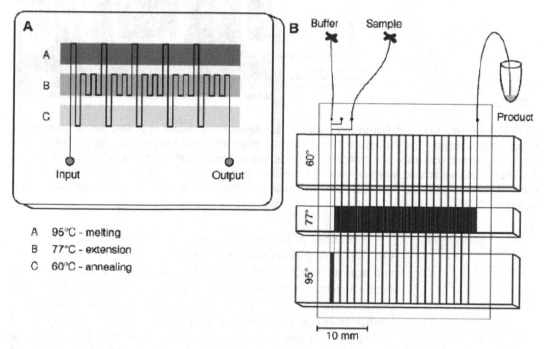

A   95°C - melting
B   77°C - extension
C   60°C - annealing

Fig. 2.7. Continuous-flow PCR on-a-chip. (a) Schematic of chip layout; (b) schematic of experimental set-up. (Reproduced with permission from Kopp et al. (120), © 1998 American Association for the Advancement of Science.)

4.2.1.2. Sequencing

A number of sequencing techniques have been developed based on DNA sequencing-by-synthesis methods. Kartalov and Quake (122) demonstrated a PDMS chip with active valves and specific surface chemistry capable of sequencing four consecutive base pairs. Electrical methods for sequence recognition may be possible, using Coulter-type devices. Nanopores could be used to resolve the DNA sequences base-by-base as the DNA molecule passes through the pore; temporal fluctuations in the current passing through the pore have been shown to relate to poly-A sequences (123). Further refinement of the technique is required but it has the potential for label-free sequencing.

4.2.2. Proteins

4.2.2.1. Immunoassays

Different variants of the standard enzyme-linked immunosorbent assay (ELISA) have been demonstrated on chip. A fast technique based on the diffusion immunoassay (DIA) using flurogenic enzymes to optically detect specific proteins was reported by Schilling et al. (124). The chip integrates cell lysis and enzyme detection and is shown in **Fig. 2.8**.

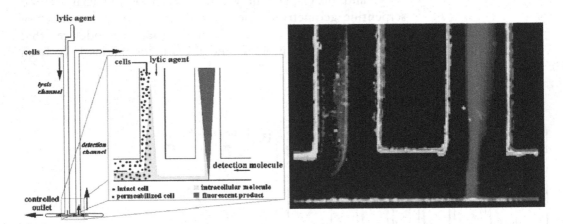

Fig. 2.8. Schematic view and fluorescence image of a diffusion immunoassay chip (DIA) for the detection of β-galactosidase enzyme. The lytic agent and the cell suspension are introduced into the same flow channel. The lytic agent diffuses into the cell suspension, lysing the cells. Intracellular components diffuse from the cell stream and a portion flow into the detection channel, where their presence is detected by the production of a fluorescent species (resorfurine) from a fluorogenic substrate. (Reproduced with permission from Schilling et al. (124), © 2002 American Chemical Society.)

A number of immunoassay systems have been demonstrated. On-chip examples include the detection of HIV-1 from infected and non-infected patients (125), the detection of cytokine tumour necrosis factor α with picomolar sensitivity (126), the integration of standard immunostrip into microflow channels for the detection of cardiac troponin I (marker for acute myocardial infarction) (127) and the use of microfabricated filters for bead-based detection of viruses (128). ELISA techniques combined with electrochemical detection have also been demonstrated by Rossier et al. (129).

5. Conclusions

In this chapter we have attempted to present a brief overview of some of the more useful and illustrative applications of microsystems technologies as applied to the biological sciences. The basic physics of microsystems was initially covered, followed by a brief description of the microfabrication technologies required to produce such devices. The majority of the microsystems technologies presented in this chapter are still in the developmental stage and chips are not yet available "off-the-shelf." It is, however, encouraging to note that the number of microfluidic chip companies is growing year-on-year and that the fabrication technologies are showing a trend towards increased simplicity. We predict that the application of microsystems-based technologies to the biological sciences will become routine within the next few years, with a diverse range of cheap, user-configurable microfluidic devices becoming widely available.

Acknowledgements

We would like to thank Hywel Morgan for his continued support, Mairi Sandison for her critical reading of the manuscript and Urban Seger for a number of useful comments regarding electroporation.

References

1. Dittrich, P. and Manz, A. (2005) Single-molecule fluorescence detection in microfluidic channels – the Holy Grail in µTAS? *Anal. Bioanal. Chem.* 382, 1771–1782.

2. http://www.caliperls.com/technology_partners/microfluidic

3. Beebe, D., Mensing, G.A. and Walker, G.M. (2002) Physics and applications of microfluidics in biology. *Annu. Rev. Biomed. Eng.* 4, 261–286.

4. Squires, T.M. and Quake, S.R. (2005) Microfluidics: fluid physics at the nanoliter scale. *Rev. Modern Phys.* 77, 977–1026.

5. Stone, H.A., Stroock, A.D. and Ajari, A. (2004) Engineering flows in small devices: microfluidics toward a lab-on-a-chip. *Annu. Rev. Fluid Mech.* 36, 381–411.

6. Schimpf, M.E., Caldwell, K. and Giddings, J.C. (2000) *Field-Flow Fractionation Handbook*. John Wiley & Sons. New York.

7. Huang, Y., et al. (1997) Introducing dielectrophoresis as a new force field for field-flow fractionation. *Biophys. J.* 73(2), 1118–1129.

8. Weigl, B.H. and Hedine, K. (2002) Lab-on-a-chip-based separation and detection technology for clinical diagnostics. *Am. Clin. Lab.* 21, 8—13.

9. Kamholz, A.E., Weigl, B.H., Finlayson, B.A. and Yager, P. (1999) Quantitative analysis of molecular interaction in a microfluidic channel: the T-sensor. *Anal. Chem.* 71, 5340–5347.

10. Baroud, C.N., Okkels, F., Ménétrier, L. and Tabeling, P. (2003) Reaction-diffusion dynamics: confrontation between theory and experiment in a microfluidic reactor. *Physical Rev. E*, 60104.

11. Hatch, A., Kamholz, A.E., Hawkins, K.R., Munson, M.S., Schilling, E.A., Weigl, B.H. and Yager, P. (2001) A rapid diffusion

immunoassay in a T-sensor. *Nature Biotechnol.* 19, 461–465.

12. Weigl, B.H. and Yager, P. (1999) Microfluidic diffusion-based separation and detection. *Science* 283, 346–347.

13. Keir, R., Igata, E., Arundell, M., Smith, W.E., Graham, D., McHugh, C. and Cooper, J.M. (2002) SERRS. In situ substrate formation and improved detection using microfluidics. *Anal. Chem.* 74, 1503–1508.

14. Verpoorte, E. (2002) Microfluidic chips for clinical and forensic analysis. *Electrophoresis* 23(5), 677–712.

15. Brody, J.P., Yager, P., Goldstein, R.E. and Austin, R.H. (1996) Biotechnology at low Reynolds numbers. *Biophys. J.* 71, 3430–3441.

16. Brody, J.P. and Yager, P. (1997) Diffusion-based extraction in a microfabricated device. *Sens. Actuators A Phys.* 58(1), 13–18.

17. Brody, J.P., Osborn, T.D., Forster, F.K. and Yager, P. (1996) A planar microfabricated fluid filter. *Sens. Actuators A Phys.* 59, 704–708.

18. Schuster, T.G., Cho, B., Keller, L.M., Takayama, S. and Smith, G.D. (2003) Isolation of motile spermatozoa from semen samples using microfluidics. *Reprod. BioMed. Online* 7(1), 75–81.

19. Morgan, H. and Green, N.G. (2003) *AC Electrokinetics: Colloids and Nanoparticles.* Research Studies Press Ltd., Baldock, UK.

20. Effenhauser, C.S., Bruin, G.J.M. and Paulus, A. (1997) Integrated chip-based capillary electrophoresis. *Electrophoresis* 18, 2203–2213.

21. Pohl, H.A. (1978) *Dielectrophoresis.* Cambridge University Press, Cambridge, UK.

22. Jones, T.B. (1995) *Electromechanics of Particles.* Cambridge University Press, Cambridge, UK.

23. Pethig, R. (1979) *Dielectric and Electronic Properties of Biological Materials.* John Wiley & Sons, Chichester, UK.

24. Pethig, R. (1996) Dielectrophoresis: using inhomogeneous AC electrical fields to separate and manipulate cells. *Critical Rev. Biotechnol.* 16(4), 331–348.

25. Yang, J., et al. (1999) Cell separation on microfabricated electrodes using dielectrophoretic/gravitational field flow fractionation. *Anal. Chem.* 71(5), 911–918.

26. Becker, F.F., Wang, X.-B., Huang, Y., Pethig R., Vykoukal J. and Gascoyne P. R.C. (1995) Separation of human breast cancer cells from blood by differential dielectric affinity. *Proc. Natl. Acad. Sci. USA* 92, 860–864.

27. Holmes, D., Green, N.G. and Morgan, H. (2003) Dielectrophoretic flow-through separation systems: comparison of experimental and numerical simulations. *IEEE Eng. Med. Biol.* 22(6), 85–90.

28. Markx, G.H., Dyda, P.A. and Pethig, R. (1996) Dielectrophoretic separation of bacteria using a conductivity gradient. *J. Biotechnol.* 51(2), 175–180.

29. Lapizco-Encinas, Simmoms B.A., Cummings E.B. and Fintschenko Y. (2004) Insulator-based dielectrophoresis for the selective concentration and separation of live bacteria in water. *Elecophoresis* 25, 1695–1704.

30. Hughes, M.P., Morgan, H. and Rixon, F.J. (2002) Measuring the dielectric properties of herpes simplex virus type I virions with dielectrophoresis. *BBA* 1571, 1–8.

31. Asbury, C.L., Diercks, A.H. and van den Engh, G. (2002) Trapping of DNA by dielectrophoresis. *Electrophoresis* 23(16), 2658–2666.

32. Washizu, M., Suzuki, S., Kurosawa, O., Nishizaka, T. and Shinohara, T. (1994) Molecular dielectrophoresis of biopolymers. *IEEE Trans. on Ind. Appl.* 30(4), 835–843.

33. Han, C. Seo, H., Lee, H., Kim, S. and Kwak, Y. (2006) Electrokinetic deposition of individual carbon nanotube onto an electrode gap. *Int. J. Precision Eng. Manufac.* 7(1), 42–46.

34. Sauer, F.A. and Schlögl, R.W. (1985) *Interactions Between Electromagnetic Fields and Cells.* (Chiabrera, A. et al., ed) Plenum, New York, pp. 203–251.

35. Archer, S., Morgan, H. and Rixon, F.J. (1999) Electrorotation studies of baby hamster kidney fibroblasts infected with herpes simplex virus type 1. *Biophys. J.* 76, 2833–2842.

36. Ziervogel, H., Glaser, R., Schadow, D. and Heymann, S. (1986) Electrorotation of lymphocytes-the influence of membrane events and nucleus. *Bioscience Reports* 6, 973–982.

37. Turcu, I. and Lucaciu, C.M. (1989) Electrorotation: a spherical shell model. *J. Phys. A: Math. Gen.* 22, 995–1003.

38. Hölzel, R. (1997) Electrorotation of single yeast cells at frequencies between 100 Hz and 1.6 GHz. *Biophys. J.* 73, 1103–1109.

39. Yang, J., Huang, Y., Wang, X.-J., Wang, X.-B., Becker, F.F. and Gascoyne, P.R.C. (1999) Dielectric properties of human leukocytes subpopulations determined by electrorotation as a cell separation criterion. *Biophys. J.* 76, 3307–3314.

40. Huang, J.P., Yu, K.W., Gu, G.Q. and Karttunen, M. (2003) Electrorotation in graded colloidal suspensions. *Phys. Rev. E* 67, 1–5.

41. Dalton, C., Goater, A.D., Burt, J.P.H. and Smith, H.V. (2004) Analysis of parasites by electrorotation. *J. Appl. Microbiol.* 96, 24–32.

42. Wang, X-B., Pethig, R. and Jones, T.B. (1992) Relationship of dielectrophoretic and electrorotational behaviour exhibited by polarized particles. *J. Phys. D: Appl. Phys.* 25, 905–912.

43. Goater, A.D. and Pethig, R. (1998) Electrorotation and dielectrophoresis. *Parasitology* 116, 177–189.

44. Gimsa, J. (2001) A comprehensive approach to electro-orientation, electrodeformation, dielectrophoresis and electrorotation of ellipsoidal particles and biological particles. *Bioelectrochemistry* 54, 23–31.

45. Madou, M. (1997) *Fundamentals of Microfabrication*. CRC Press, Boca Raton.

46. Kovacs, G. (1998) *Micromachined Transducers Sourcebook*. WCB-McGraw-Hill, Boston.

47. Becker, H. and Locascio, L.E. (2002) Polymer microfluidic devices. *Tanlanta* 56(2), 267–287.

48. Xia, Y. and Whitesides, G.M. (1998) Soft Lithography. *Ann. Rev. Mater. Sci.* 28, 153–184.

49. Wu, H., Odom, T.W., Chui, D.T. and Whitesides, G.M. (2003) Fabrication of complex three-dimensional microchannel systems in PDMS. *J. Am. Chem. Soc.* 125, 554–559.

50. Jo, B., van Lerberghe, L.M., Motsegood, K.M. and Beebe, D.J. (2000) Three-dimensional micro-channel fabrication in polydimethylsiloxane (PDMS) elastomer. *J. Microelectromech. Syst.* 9(1), 76–81.

51. Moorcroft, M.J., Meuleman, W.R.A., Latham, S.G., Nicholls, T.J., Egeland, R.D. and Southern, E.M. (2005) In situ oligonucleotide synthesis on poly(dimethylsiloxane): a flexible substrate for microarray fabrication. *Nucleic Acids Res.* 33(8), e75.

52. Gerlach, A., Knebel, G., Guber, A.E. Heckele, M., Herrmann, D., Muslija, A. and Schaller, T. (2002) High-density plastic microfluidic platforms for capillary electrophoresis separation and high-throughput screening. *Sensors Mater.* 14(3), 119—128.

53. Wang, J., Pumera, M., Chatrathi, M.P., Escarpa, A., Konrad, R., Griebel, A., Dörner, W. and Löwe, H. Towards disposable lab-on-a-chip: Poly(methylmethacrylate) microchip electrophoresis device with electrochemical detection. *Electrophoresis* 23(4), 596–601.

54. Chou, S.Y., Krauss, P.R. and Renstrom, P.J. (1996) Imprint lithography with 25-Nanometer resolution. *Science* 272(5258), 85–87.

55. Eldada, L. and Shacklette, L.W. (2000) Advances in polymer integrated optics. *Selected Topics in Quantum Electronics, IEEE J.* 6(1), 54–68.

56. Heckele, M., Bacher, W. and Muller, K.D. (1998) Hot embossing – The molding technique for plastic microstructures. *Microsyst. Technol.* 4, 122–124.

57. De, S.K., Aluru, N.R., Johnson, B., Crone, W.C., Beebe, D.J. and Moore, J. (2002) Equilibrium swelling and kinetics of pH-responsive hydrogels: models, experiments, and simulations. *J. Microelectromech. Syst.* 11, 544–555.

58. Hoffman, J., Plotner, M., Kuckling, D. and Fischer, W. (1999) Photopatterning of thermally sensitive hydrogels useful for microactuators. *Sens. Actuators B-Chemical* 77, 139–144.

59. Suzuki, A. and Tanaka, T. (1990) Phase transition in polymer gels induced by visable light. *Nature* 346, 345–347.

60. Kataoka, K., Miyazaki, H., Bunya, M., Okano, T. and Sakurai, Y. (1998) Totally synthetic polymer gels responding to external glucose concentration: their preparation and application to on-off regulation of insulin release. *J. Amer. Chem. Soc.* 120, 12694–12695.

61. Eddington, D.T. and Beebe, D.J. (2004) A valved responsive hydrogel microdispensing device with integrated pressure source. *J. Microelectromech. Syst.* 13(4), 586–593.

62. Bertsch, A., Lorenz, H. and Renaud, P. (1999) 3D microfabrication by combining microstereolithography and thick resist UV lithography. *Sens. Actuators A-Physical* 73, 14–23.

63. Belloy, E., Thurre, S., Walckiers, E., Sayah, A. and Gijs, M.A.M. (2000) The introduction of powder blasting for sensor and microsystem applications. *Sens. Actuators A-Physical* 84, 330–337.

64. Jones, T.B. (2002) On the relationship of dielectrophoresis and electrowetting. *Langmuir* 18, 4437–4443.

65. Schwartz, J.A., Vykoukal, J.V. and Gascoyne, P.R.C. (2004) Droplet-based chemistry on a programmable micro-chip. *Lab Chip* 4, 11–17.

66. Stemme – droplet based chemical reactors.

67. Voskerician, G., Shive, M.S., Shawgo, R.S., von Recum, H., Anderson, J.M., Cima, M.J. and Langer, R. (2003) Biocompatibility and biofouling of MEMS drug delivery device. *Biomaterials* 24, 1959–1967.

68. Dittrich, P.S. and Manz, A. (2006) Lab-on-a-chip: microfluidics in drug discovery. *Nat. Rev. Drug Disc.* 5, 210–218.

69. Abhyankar, V.V., Bittner, G.N., Causey, J.A., Kamp, T.J. and Beebe, D.J. (2003) in *Micro Total Analysis Systems* 17–20 (Squaw Valley).

70. Heuschkel, M.O., Guérin, L., Buisson, B., Bertrand, D. and Renaud, P. (1998) Buried microchannels in photopolymer for delivering of solutions to neurons in a network. *Sens. Actuators B-Chemical* 48, 356–361.

71. King, K.R., Terai, H., Wang, C.C., Vacanti, J.P. and Borenstein, J.T. (2001) in *Micro Total Analysis Systems* 247–249 (Monterey).

72. Chin, V.I., Taupin, P., Sanga, S., Scheel, J., Gage, F.H. and Bhatia, S.N. (2004) Microfabricated platform for studying stem cell fates. *Biotechnol. Bioeng.* 88, 399–415.

73. Lee, P.J., Hung, P.J., Rao, V.M. and Lee, L.P. (2005) Nanoliter scale microbioreactor array for quantitative cell biology. *Biotechnol. Bioeng.* 94(1), 5–14.

74. Martin, K. et al. (2003) Generation of larger numbers of separated microbial populations by cultivation in segmented-flow microdevices. *Lab Chip* 3, 202–207.

75. Kojima, K., Moriguchi, H., Hattori, A., Kaneko, T. and Yasuda, K. (2003) Two-dimensional network formation of cardiac myocytes in agar microculture chip with 1480 nm infrared laser photothermal etching. *Lab Chip* 3, 292–296.

76. Borkholder, D.A., Opris, I.E., Maluf, N.I. and Kovacs, G.T.A. (1996) in *18th Annual International Conference of the IEEE in Medicine and Biology Society*, 106–107 (Amsterdam).

77. Heuschkel, M., Fejtl, M., Raggenbass, M., Betrand, D. and Renaud, P. (2002) A three-dimensional multi-electrode array for multi-site stimulation and recording in acute brain slices. *J. Neurosci. Methods* 114, 135–148.

78. Stett, A. et al. (2003) Biological applications of microelectrode arrays in drug discovery and basic research. *Anal. Bioanal. Chem.* 377, 486–495.

79. Brischwein, M., et al. (2004) Funtional cellular assays with multiparametric silicon chips. *Lab Chip* 3, 2324–240.

80. Yang, M., Li, C. and Yang, J. (2002) Cell docking and on-chip monitoring of cellular reactions with a controlled concentration gradient on a microfluidic device. *Anal. Chem.* 74, 3991–4001.

81. Mrksich, M. and Whitesides, G.M. (1996) Using self-assembled monolayers to understand the interactions of man-made surfaces with proteins and cells. *Proc. Natl. Acad. Sci. USA* 93, 10775–10778.

82. Chen, S.C., Mrksich, M., Huang, S., Whitesides, G.M. and Ingber, D.E. (1998) Micropatterned surfaces for control of cell shape, position, and function. *Biotechnol. Prog.* 14, 356–363.

83. Takayama, S., Ostuni, E., LeDuc, P., Naruse, K., Ingber, D.E. and Whitesides, G.M. (2001) Subcellular positioning of small molecules. *Nature* 411, 1016.

84. Takayama, S., Ostuni, E., LeDuc, P., Naruse, K., Ingber, D.E. and Whitesides, G.M. (2003) Selective chemical treatment of cellular microdomains using multiple laminar streams. *Chem. Biol.* 10, 123–130.

85. Weaver, J.C. and Chizmadzeh, Y.A. (1996) Theory of electroporation: a review. *Bioelectrochem. Bioenergetics* 41, 135–160.

86. Huang, Y. and Rubinsky, B. (2001) Microfabricated electroporation chip for single cell membrane permeabilization. *Sens. Actuators A-Physical* 89, 242–249.

87. Fox, M., Esveld, D., Valero, A., Luttge, R., Mastwijk, H., Bartels, P., van den Berg, A. and Boom, R. (2006) Electroporation of cells in microfluidic devices: a review. *Anal. Bioanal. Chem.* 385(3), 474–485.

88. Beebe, S.J., et al. (2003) Diverse effects of nano pulsed electric fields on cells and tissues. *DNA Cell Biol.* 22, 785–796.

89. Meldrum, R.A., Bowl, M., Ong, S.B. and Richardson, S. (1999) Optimisation of electroporation for biochemical experiments in live cells. *Biochem. Biophys. Res. Commun.* 256, 235–239.

90. Seger, U., Gawad, S., Johann, R., Bertsch, A. and Renaud, P. (2004) Cell immersion and cell dipping in microfluidic devices. *Lab Chip* 4, 148–151.

91. McClain, M.A., Culbertson, C.T., Jacobson, S.C. and Ramsey, J.M. (2001) in *Micro Total Analysis System* 301–302 (Monterey).

92. Lu, H., Gaudet, S., Sorger, P.K., Schmidt, M.A. and Jensen, K.F. (2003) in *Micro Total Analysis Systems*, 773–776 (Squaw Valley).

93. Ocrvirk, G., et al. (1998) in *Micro Total Analysis Systems* 203—206 (Banff).

94. Shapiro, H.M. (2003) *Practical Flow Cytometry*. John Wiley & Sons, Inc. New York.

95. Schrum, D.P., Culbertson, C.T., Jacobson, S.C. and Ramsey, J.M. (1999) Microchip flow cytometry using electrokinetic focusing. *Anal. Chem.* 71, 4173–4177.

96. Becker, C.K., Parker, M.D. and Hechinger, M.K. (2003) in *ISAC XXI*.

97. Wolff, A., Larsen, U.D., Blankenstein, G., Philip, J. and Telleman, P. (1998) in *Micro Total Analysis Systems*, 73–76 (Banff).

98. Altendorf, E., et al. (1998) in *Micro Total Analysis Systems*, 73–76 (Banff).

99. Fu, A.Y., Spence, C., Scherer, A., Arnold, F.H. and Quake, S.R. (1999) A microfabricated fluorescence-activated cell sorter. *Nat. Biotechnol.* 17, 1109–1111.

100. Kruger, J., et al. (2002) Development of a microfluidic device for fluorescence activated cell sorting. *J. Micromech. Microeng.* 12, 486–494.

101. Emmelkamp, J., DaCosta, R., Andersson, H. and van den Berg, A. (2003) in *Micro Total Analysis Systems*, 85–88 (Squaw Valley).

102. Ashkin, A. (1970) Acceleration and trapping of particles by radiation pressure. *Phys. Rev. Lett.* 24(4), 156.

103. Molloy, J.E. and Padgett, M.J. (2002) Lights, action: optical tweezers. *Contemporary Phys.* 43(4), 241–258.

104. Leach, J., Sinclair, G., Jordan, P., Courtial, J., Padgett, M., Cooper, J. and Laczik, Z. (2004) 3D manipulation of particles into crystal structures using holographic optical tweezers. *Optics Express* 12(1), 220–226.

105. Larsen, U.D., Blankenstein, G. and Ostergaard, S. (1997) in *Proceedings of Transducers 1997*, 1319—1322 (Chicago, USA).

106. Koch, M., Evans, A.G.R. and Brunnschweiler, A. (1998) in *Proceedings Micromechanics Europe 1998*, 155–158 (Ulvik, Norway).

107. Koch, M., Evans, A.G.R. and Brunnschweiler, A. (2000) *Microfluidic Technology and Applications*. Research Studies Press Ltd, Baldock, UK.

108. Fuller, C.K., Hamilton, J., Ackler, H. and Gascoyne, P.R.C. (2000) *Microfabricated Multi-Frequency Particle Impedance Characterization System*, in Micro Total Analysis Systems 2000, Kluwer, Enschede, Netherland, pp. 265–268.

109. Larsen, U.D., Norring, H. and Telleman, P. (2000) in Micro Total Analysis Systems 2000, Kluwer, Enschede, Netherland, pp. 103–106.

110. Gawad, S., et al. (2000), in IEEE-EMBS Conference on Microtechnologies in Medicine & Biology, Lyon, France, pp. 297–301.

111. Benazzi, G., Holmes, D., Mowlem, M. and Morgan, H. (2006) Discrimination and analysis of phytoplankton using a microfabricated flow-cytometer. (Submitted *Cytometry – Part A*.)

112. Gawad, S., Tao, S., Holmes, D. and Morgan, H. Broadband dielectric cytometry using nanoporous electrodes and maximum length sequences. Nanotech 2005, Montreux, Switzerland.

113. Holmes, D., Morgan, H. and Green, N.G. (2006) Flow cytometry in a microchip: combining optical detection with dielectrophoretic focusing. *Biosensors and Bioelectronics* 21, 1622–1630.

114. Holmes, D., Sandison, M.E., Green, N.G. and Morgan, H. (2005) On-chip high-speed sorting of micron-sized particles for high-throughput analysis. *IEE Proc. Nanobiotechnol.* 152(4), 129–135.

115. Fuhr, G., et al. (1994) Radiofrequency microtools for particle and live cell manipulation. *Naturwissenschaften* 81(12), 528–535.

116. Schrum, D.P., et al., (1999) Microchip flow cytometry using electrokinetic focusing. *Anal. Chem.* 71(19), 4173–4177.

117. Ceriotti, L., de Rooij, N.F. and Verpoorte, E. (2002) An integrated fritless column for on-chip capillary electrochromatography with conventional stationary phases. *Anal. Chem.* 74, 639–647.

118. Ros, A., et al. (2002) Protein purification by off-gel electrophoresis. *Proteomics* 2, 151–156.

119. Macounova, K., Carbrera, C.R., Holl, M. and Yager, P. (2000) Generation of natural pH gradients in microfluidic channels for use in isoelectric focusing. *Anal. Chem.* 72, 3745–3751.

120. Kopp, M.U., de Mello, A.J. and Manz, A. (1998) Chemical amplification: continous-flow PCR on a chip. *Science* 280, 1046–1048.

121. Waters, L.C., et al. (1998) Microchip device for cell lysis, multiplex PCR amplification, and electrophoretic sizing. *Anal. Chem.* 70, 158–162.

122. Braslavsky, I., Hebert, B., Kartalov, E. and Stephen R. Quake (2003) Sequence information can be obtained from single DNA molecules. *Proc. Natl. Acad. Sci. USA* 100, 3960–3964.

123. Bayley, H. and Martin, C.R. (2000) Resistive-pulse sensing from microbes to molecules. *Chem. Rev.* 100, 2575–2594.

124. Schilling, E.A., Kamholz, A.E. and Yager, P. (2002) Cell lysis and protein extraction in a

microfluidic device with detection by a fluorogenic enzyme assay. *Anal. Chem.* 74, 1798–1804.

125. Sia, S.K., Linder, V., Parviz, B.A., Siegel, A. and Whitesides, G.M. (2004) An integrated approach to a portable and low-cost immunoassay for resource-poor settings. *Angew. Chem.* 43, 498–502.

126. Cesaro-Tadic, S., Dernick, G., Juncker, D., Buurman, G., Kropshofer, H., Michel, B., Fattinger, C. and Delamarche, E. (2004) High-sensitivity immunoassays for tumor necrosis factor α using microfluidic systems. *Lab Chip* 4, 563–569.

127. Cho, J.H., Han, S.M., Paek, E.H., Cho, I.H. and Paek, S.H. (2006) Plastic ELISA-on-a-chip based on sequential cross-flow chromatography. *Anal. Chem.* 78(3), 793–800.

128. Liu, W.T., Zhu, L., Qin, Q.W., Zhang, Q., Feng, H. and Ang, S. (2005) Microfluidic device as a new platform for immunofluorescent detection of viruses. *Lab Chip* 5, 1327–1330.

129. Rossier, J.S. and Girault, H.H. (2001) Enzyme linked immunosorbent assay on a microchip with electrochemical detection. *Lab. Chip.* 1, 153–157.

Chapter 3

Rapid Prototyping of Microstructures by Soft Lithography for Biotechnology

Daniel B. Wolfe, Dong Qin, and George M. Whitesides

Abstract

This chapter describes the methods and specific procedures used to fabricate microstructures by soft lithography. These techniques are useful for the prototyping of devices useful for applications in biotechnology. Fabrication by soft lithography does not require specialized or expensive equipment; the materials and facilities necessary are found commonly in biological and chemical laboratories in both academia and industry. The combination of the fact that the materials are low-cost and that the time from design to prototype device can be short (< 24 h) makes it possible to use and to screen rapidly devices that also can be disposable. Here we describe the procedures for fabricating microstructures with lateral dimensions as small as 1 μm. These types of microstructures are useful for microfluidic devices, cell-based assays, and bioengineered surfaces.

Key words: Soft lithography, poly(dimethylsiloxane), microfluidics, microfabrication, rapid prototyping, cell-based assays.

1. Introduction

Most research in microfabrication is focused on applications in microelectronics. Applications in biotechnology are, however, rapidly emerging: these include tools for cell-based assays (1–5), molecular (DNA sequencing on microarrays) (6–12) and clinical (enzyme-linked immunosorbant assay (ELISA) on microchips) diagnostics (13–17), drug discovery (1–5), and chemical and biological defense (microchip-based sensors of chemicals and pathogens) (18–24).

There are four general steps in the process of microfabrication: (1) fabrication of a master (i.e., the pattern used to make replicas), (2) replication of the master, (3) transfer of the pattern in the

M.P. Hughes, K.F. Hoettges (eds.), *Microengineering in Biotechnology*, Methods in Molecular Biology 583,
DOI 10.1007/978-1-60327-106-6_3, © Humana Press, a part of Springer Science+Business Media, LLC 2010

replica into functional materials (e.g., polymers, metals, ceramics, or biologics), and (4) registration of a master (the original one or a different one) with the patterned material in Step 3 to form multi-level structures. The materials patterned by the techniques used in the microelectronics industry (e.g., photolithography and electron-beam lithography) are limited to photosensitive polymers. These techniques can be used to pattern DNA (10–12), but they have limited capability to pattern other biologically relevant materials (e.g., proteins and cells) directly.

Soft lithography is set of techniques that provides alternatives to photolithography for Steps 2–4; these techniques are useful for the prototyping of devices useful in biology, chemistry, and physics (1, 25–29). They use a topologically patterned stamp (or mold) to transfer a pattern to a substrate. There are two types of soft lithography: (1) the replication of a pattern defined in a *soft* (elastomeric) stamp into organic materials or onto the surface of metals, ceramics, and semiconductors and (2) the replication of a pattern defined in a *hard* (rigid) stamp into a thin layer of *soft* organic molecules (e.g., thermoplastic polymers). In this chapter we discuss the use of the first type of soft lithography in the fabrication of microstructures useful in biology.

The types of microstructures useful for biological applications have requirements different from those in microelectronics. Microfluidic systems, for example, are relatively simple in design and have design rules that accept large features (i.e., feature sizes of 100 µm and areas of >5 cm^2). It is also useful to be able to try a large number of designs of microfluidic systems and bioanalytical tools rapidly and at low cost for each device. Soft lithography also makes it possible for researchers to fabricate prototype devices rapidly (i.e., from design to prototype in 24–48 h) using equipment found commonly in most scientific laboratories. The capabilities of soft lithography overlap well with the requirements of the fabrication of microdevices needed in biology and biochemistry.

A number of reviews and papers describe the types of biological experiments that have been performed using devices fabricated by soft lithography (1, 25–28, 30, 31). Here we describe how to fabricate membranes with holes, microfluidic channels, and engineered surfaces by using rapid prototyping and soft lithography.

2. Materials

Note: The companies listed below are not the sole providers of these materials.

2.1. Fabrication of Masters

1. Computer drawing software (e.g., Adobe Photoshop, Macromedia Freehand, WieWeb Clewin)

2.1.1. Fabrication of Transparency-Based Photomasks

2.1.2. Contact Photolithography Using Transparency-Based Photomasks

1. Contact mask aligner equipped with a mercury-arc lamp (found in most cleanroom facilities)

2.1.3. Microscope Projection Photolithography (MPP)

1. Upright microscope equipped with a mercury-arc lamp

2.1.4. Microlens Array Photolithography (MAP)

1. Standard overhead projector equipped with a broadband light bulb
2. Aqueous hydrogen peroxide (20%$_{vol}$) (*see* **Note 1**)
3. Concentrated sulfuric acid (*see* **Note 1**)

2.1.5. Materials Common to All of the Photolithographic Techniques

1. Substrates – silicon wafers (test grade, <100>), glass slides, gold-coated glass slides (Platypus Technologies, Madison, WI). *Note: Cracked silicon wafers can be very sharp. Please use caution when handling cracked wafers.*
2. Cleaning solution – trichloroethylene (TCE), acetone, and methanol (Aldrich Chemical Co., St. Louis, MO)
3. Primer – hexamethyldisilazane (HMDS) (Aldrich Chemical Co., St. Louis, MO)
4. Positive photoresist – Microposit 1813, 1805 (Shipley Co., Inc., Marlborough, MA)
5. Negative photoresist – SU-8 50 (Microchem Corp., Newton, MA)
6. Developer – Microposit 351 (Shipley Co., Inc., Malborough, MA); Propylene glycol methyl ether acetate (Aldrich Chemical Co., St. Louis, MO)

2.2. Fabrication of Poly(dimethylsiloxane) (PDMS) Replicas

1. Surface treatment – Tridecafluoro-1,1,2,2-tetrahydrooctyl-1-trichlorosilane ($CF_3(CF_2)_6(CH_2)_2SiCl_3$) (United Chemical Technology, Bristol, PA)
2. Glass vacuum dessicator (VWR Scientific, West Chester, PA)
3. Elastomeric materials – Poly(dimethylsiloxane) – Sylgard 184 Silicon Elastomer Kit (Dow Corning, Highland, MI)

2.2.1. PDMS Membranes

1. Spin coater capable of spinning at 500–5,000 rpm

2.3. Fabrication of Microfluidic Channels

1. 16-gauge needles (VWR Scientific, West Chester, PA)
2. Metal file
3. Sand paper
4. Sharp tweezers
5. Oxygen plasma cleaner (Harrick Scientific Corporation, Ossining, NY)
6. Polyethylene tubing (PE-60, VWR Scientific, West Chester, PA)

2.4. Fabrication of Engineered Surfaces

2.4.1. Microcontact Printing of SAMs to Pattern Cells

1. 1-octadecanethiol (Aldrich Chemical Company, St. Louis, MO)
2. Cotton swabs
3. Tri(ethylene glycol)-terminated undecanethiol (Prochima, Poland; Shearwater Polymers, Huntsville, AL)

2.4.2. Microcontact Printing of Reactive SAMs to Pattern Ligands

1. Hexa(ethylene glycol)-terminated undecanethiol (Prochima, Poland; Shearwater Polymers, Huntsville, AL)
2. 1-Ethyl-3-(dimethylamino)propylcarbodiimide (EDC) (Aldrich Chemical Company, St. Louis, MO)
3. Pentafluorophenol (Aldrich Chemical Company, St. Louis, MO)
4. Biotin cadaverine (Molecular Probes, Inc., Eugene, OR)

---

3. Methods

This section details the steps and procedures used to take a design to a functional prototype by rapid prototyping and soft lithography. The organization of this section follows that of the outline shown in **Fig. 3.1**.

3.1. Fabrication of Masters

3.1.1. Fabrication of Transparency-Based Photomasks

The master contains the original pattern to be transferred into the elastomeric stamp by molding. The minimum size of the features necessary in the master is dictated by the specific application of the device. The master in photolithography – a fabrication technique used commonly in the microelectronics industry – is a photomask. Photoresist-based replicas generated by this technique are used as the masters for the fabrication of elastomeric stamps for soft lithography. The photomasks are generated by laser or electron-beam writing on photoresist that is coated on a chrome-coated sheet of float glass or quartz. This process exposes areas of the chrome; these regions of chrome are removed by wet-chemical etching.

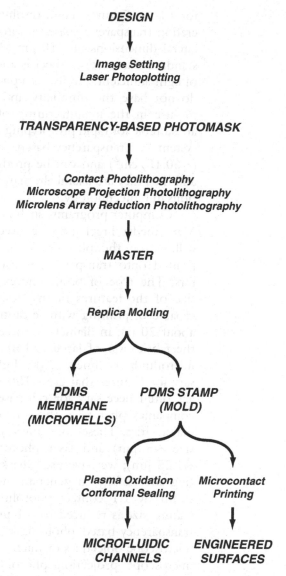

Fig. 3.1. Diagram of the flow of the process of microfabrication of elastomeric membranes with holes, microfluidic channels, and engineered surfaces by soft lithography.

Electron-beam and laser writing are capable of generating small features (e.g., <100 nm), but are expensive ($\sim$\$50/cm^2) and slow (>4 h to pattern 5 cm^2). Many applications in biology do not require such resolution, and appropriate masks can, therefore, be prepared using low-cost methods and materials. For example, the range of sizes of the smallest lateral dimension of the features used for microfluidic channels are 50–500 μm, for single-cell assays are 5–50 μm, and for subcellular assays are 1–5 μm (1, 27, 30). There are few (if any) common applications in biotechnology for <1-μm features. High-resolution printers are capable of generating transparency-based photomasks containing features with lateral dimensions of >10 μm (32, 33). Black ink printed on a standard transparency sheet is able to attenuate the transmission of light sufficiently to use as a photomask. Although these masks do not have the durability and dimensional stability required for use in the manufacturing of microelectronic devices, they are suitable for rapid prototyping of bioanalytical and microfluidic systems. Transparency-based photomasks are inexpensive (~$0.15/cm^2) and can be produced rapidly (>200 cm^2/min), and thus they are useful alternatives to chrome masks for applications in biotechnology.

Computer programs, such as Adobe Illustrator, Clewin, and Macromedia Freehand, are used to "draw" the patterns that will be on the photomask (*see* **Note 2**). These designs are printed onto transparencies using commercially available printers. The type of printer necessary depends on the minimum size of the features in the design. Standard laser printers are capable of printing with resolution of ~1,200 dpi (each dot is about 20 μm in diameter); these photomasks are acceptable for the fabrication of large (>150 μm) features, such as those in microfluidic channels (33). High-resolution printers are necessary for features that are <150 μm in width; such feature sizes are often necessary for cell-based assays. Commercial printing companies phat have these types of printers operate in most major cities. These companies use imagesetters (5,080 dpi; dot size ~5 μm) and laser photoplotters (20,000 dpi; dot size ~1.25 μm); we have used masks generated by laser photoplotting successfully to generate high-quality features with 10-μm width by 1:1 contact photolithography (32). This minimum feature size is reduced to ~1 μm by projecting the patterns in transparency-based photomasks through reducing optics (e.g., microscope objectives or microlens arrays) in techniques such as microscope projection photolithography (34) and microlens projection photolithography (35–39).

3.1.2. Contact Photolithography Using Transparency-Based Photomasks

Standard photolithography using transparency masks is used to fabricate masters that have a single pattern with feature sizes of ≥10 μm. The maximum area of the master is dictated by the diameter of the aperture of the light source; the maximum diameter is typically 75–100 mm (3–4 inches). Masters with these dimensions and feature sizes are useful for the fabrication of microfluidic channels and elastomeric membranes with holes.

The step-by-step procedure for photolithography includes the preparation of the substrate, the exposure of the photoresist, and the removal of the unwanted photoresist (**Fig. 3.2**). A substrate is coated with a thin layer of photosensitive polymer (photoresist) by

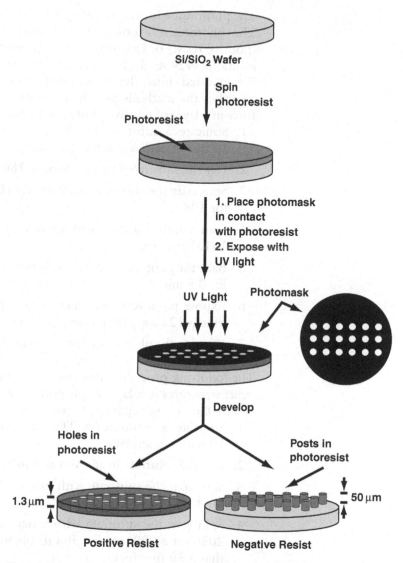

Fig. 3.2. Schematic diagram of the process of contact photolithography.

spin coating (*see* **Note 3**). The thickness of these layers can range from a few nanometers up to a millimeter depending on the type of resist that is used and the conditions used in spinning; SU-8 (Microchem Corp, Newton, MA) is used commonly for layers that are more than 10-μm thick. The transparency-based photomask is placed in contact with the photoresist-coated substrate such that the side with the ink is in contact with the surface of the photoresist (*see* **Note 4**). A transparent glass plate is placed on top of the photomask. Pressure is applied between the photomask and the substrate by clamping them between two rigid plates (e.g., metal plates with a hole in the center to allow light to pass) to press the photomask into good contact with the photoresist.

The photoresist is exposed using a UV-light source (*see* **Note 3**). The photoresist exposed to UV light becomes either more (positive resist) or less (negative resist) soluble in the developing solution (*see* **Note 3**). In either case, the pattern of the photomask is replicated into the photoresist. The following procedure describes the methods used to generate a patterned relief structures in a layer of positive photoresist with thickness of 1.3 μm:

1. Sonicate a substrate in TCE, acetone, and methanol for 10 min in each solvent, sequentially.

2. Dry the substrate in an air oven at 180°C for 10 min.

3. Spin coat the substrate with primer (HMDS) at 4,000 rpm for 40 s.

4. Spin coat the substrate with a positive photoresist (Microposit 1813) at 4,000 rpm for 40 s.

5. Bake the photoresist-coated substrate on a hot plate at 105°C for 3.5 min.

6. Expose photoresist through the photomask. The exposure time is 12 s for a lamp intensity of 10 mJ cm^{-2}s^{-1} at 405 nm.

7. Develop the photoresist for 1 min in dilute developer (Microposit 351:H$_2$O = 1:5 vol/vol)

The following procedure describes the fabrication of a patterned relief structures in a layer of photoresist with thickness of greater than 10 μm using negative photoresists:

1. Sonicate a substrate in TCE, acetone, and methanol for 10 min, sequentially.

2. Dry the substrate in an oven at 180°C for 10 min.

3. Spin coat the substrate with primer (HMDS) at 4,000 rpm for 40 s.

4. Spin coat the substrate with a negative photoresist (SU-8-2050) at 3,000 rpm for 30 s to obtain a layer of photoresist that is 50-μm thick.

5. Bake the photoresist-coated substrate on a hot plate at 65°C for 3 min and then at 95°C for 6 min.

6. Expose photoresist through the photomask. The exposure time is 40 s for a lamp intensity of 10 mJ cm^{-2}s^{-1} at 365 nm.

7. Develop the photoresist for 6 min in propylene glycol methyl ether acetate.

3.1.3. Microscope Projection Lithography (MPP)

Microscope projection photolithography (MPP) is useful for generating features smaller than those possible using contact photolithography with transparency photomasks (i.e., 1–10 μm in lateral dimensions), but the technique is limited to patterning small areas ($\sim 4 \times 10^4$ μm^2) per exposure (34). Photoresist-based masters with these dimensions and feature sizes are useful for fabricating microwells and elastomeric membranes with holes (40).

Microscope projection photolithography (MPP) uses a standard upright microscope equipped with a mercury-arc lamp to reduce the size of features defined in a transparency photomask by up to 25 × (**Fig. 3.3**). Mercury-arc lamps emit UV light at the wavelengths necessary for the exposure of photoresist (i.e., 436 nm). These light sources are commonly used for fluorescence microscopy. A transparency photomask is placed before the objective in a location that is a conjugate image plane to that of the surface of the substrate (*see* **Note 5**). The reduction factor of the features on the mask depends on the magnification power of the objective (*see* **Note 5**). Typical exposure times ranged from 5 to 20 s (80 W Mercury Lamp) depending on the thickness of the resist (*see* **Note 6**). Microscope projection lithography can produce features with widths of 1 μm and an edge roughness of 0.2 μm. A drawback of the technique is that it is limited to the patterning of areas of ~40,000 μm^2 (0.4 mm^2) for a single exposure. Rastering of the stage with subsequent exposures permits the fabrication of arrays of features. This technique is also useful for the registration of multiple exposures with high accuracy. **Figure 3.3d** shows examples of features patterned in photoresist by this technique. A comprehensive procedure is described as follows (*see* **Note 6**):

1. A transparent photomask is placed in the microscope in a location between the light source and the back aperture of the objective that is the conjugate image plane of the surface of the photoresist of the microscope.

2. A photoresist-coated substrate is prepared as described previously. It is placed on the microscope stage underneath the microscope objective (*see* **Note 7**).

3. A high neutral-density filter is placed in front of the light source to prevent accidental exposure of the photoresist while focusing.

4. The neutral-density filter is removed to expose the resist.

5. The photoresist-coated wafer is rastered to an unexposed area and Steps 3 and 4 are repeated to fabricate arrays of the features.

6. The exposed resist is developed using appropriate developer solutions (*see* **Note 3**).

3.1.4. Microlens Array Projection Lithography (MAP)

Microlens projection photolithography (MAP) is a technique that is capable of patterning arrays of the same feature over areas larger than 1 cm^2 in a single exposure (35–39). Masters with these types of features are useful for the fabrication of microwells and elastomeric membranes with holes over areas larger than those possible with microscope projection photolithography.

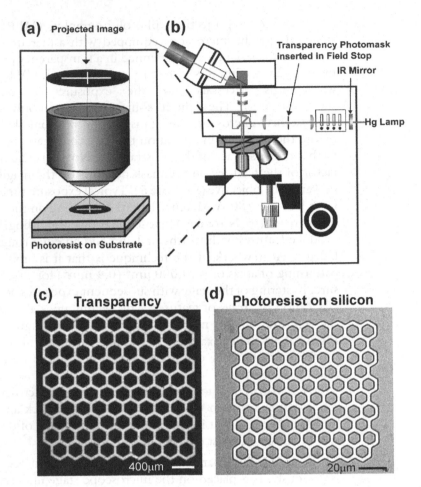

(a) Projected Image

Photoresist on Substrate

(b) Transparency Photomask inserted in Field Stop

IR Mirror

Hg Lamp

(c) Transparency

400µm

(d) Photoresist on silicon

20µm

Fig. 3.3. (**a and b**) Schematic diagram of microscope projection photolithography. (**c**) Optical micrograph of transparency mask. (**d**) Optical micrograph of the pattern in (**c**) generated in photoresist by MPP. Images are reprinted with permission from Love et al. (34) © 2001 American Chemical Society.

Microlens array projection lithography (MAP) uses microlenses (3–100 µm in diameter) to reduce the size of a large image onto the photoresist by a factor of ~800 × . The minimum feature size that can be patterned by this technique is ~600 nm. **Figure 3.4a** diagrams the process for the technique. A transparency mask is placed on the surface of a standard overhead projector. The projection optics are removed from the projector and replaced with the microlenses. The microlenses are made of photoresist posts that have been melted and reshaped into hemispheres on the surface of a glass slide (**Fig. 3.5**). The back of the microlens array is coated with a layer of poly(dimethylsiloxane) (PDMS) to allow for uniform contact of the photoresist-coated wafer with the lens array. The thickness of the PDMS layer is set at the focal length of the lenses; the focal length is determined by the diameter and the height of the lens. This technique is uniquely suited for the

(a)

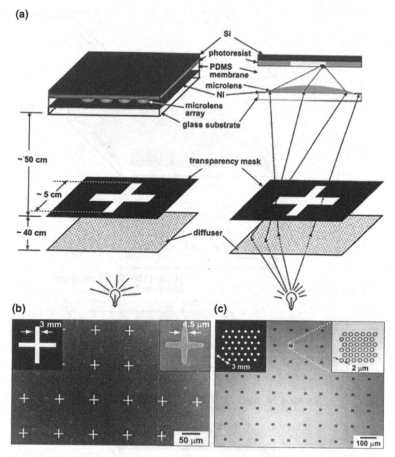

Fig. 3.4. **(a)** Schematic diagram of microlens array projection photolithography. **(b and c)** Scanning electron micrographs of arrays of features generated by the technique. The schematic diagram is reprinted from reference 36 with permission from the American Chemical Society. The images are reprinted from Wu et al., Generation of Chrome Masks with Micrometer-Scale Features using Microlens Lithography. *Adv. Mater.* 14, 1213–1216, 2002. Copyright Wiley VCH Verlag GmbH & Co. KGaA. Reproduced with permission.

preparation of repetitive structures because it does so in a single exposure with minimal distortion (*see* **Note 8**). **Figure 3.4b** and **c** show examples of the types of features that have been fabricated by this technique. A detailed procedure for microlens projection photolithography using an array of 20-μm-diameter lenses is given below:

1. A cleaning solution (piranha-etch solution) is prepared by *slowly* adding aqueous hydrogen peroxide ($20\%_{vol}$) to concentrated sulfuric acid at room temperature to reach a 1:2 ratio by volume. *Note: The ingredients of this solution and the solution itself can cause severe burns, and contact with organic materials or solvents can result in explosions. See Note 1 for a discussion of the dangers of the piranha solution and proper handling procedures.*

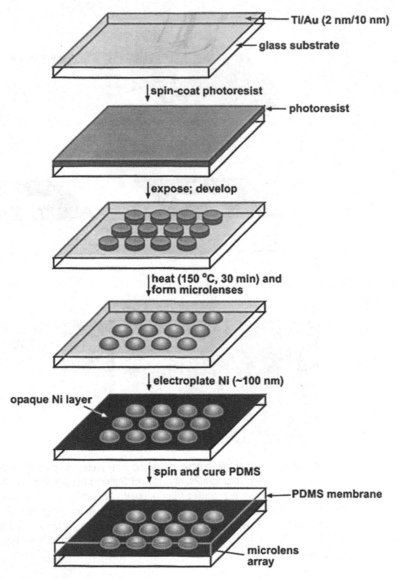

Fig. 3.5. Schematic diagram of the procedure used to make an array of microlenses. Scheme is reprinted permission from Wu et al. (36) © 2002 American Chemical Society.

2. Glass slides are immersed in the piranha-etch solution for 24 h (*see* **Note 8**).

3. A thin layer of titanium (2 nm), followed by gold (10 nm), is deposited by physical vapor deposition.

4. A 2-μm-thick layer of positive photoresist (Shipley S1818; Shipley Corporation) is spin coated (3,000 rpm) onto the gold-coated glass substrate.

5. A transparency mask with a 1 × 1 cm array of 20-μm-diameter posts is prepared as described previously.

6. The photoresist is exposed by standard photolithography for 12 s and developed in dilute developer (Microposit 351:water = 5:1) for 1 min.

7. The posts are melted and reshaped into hemispheres by heating the array on a hot plate at 150°C for 30 min.

8. The focal length, f, of the microlenses is calculated using the following equation (36):

$$f = ((D/2)^2 + s^2)/2s(n_{\text{lens}} - n_{\text{PDMS}})$$

where D is the diameter of the lens (μm), s is the height of the lens (μm), n_{lens} is the refractive index of the photoresist (1.59), and n_{PDMS} is the refractive index of PDMS (1.405).

9. A fresh mixture of PDMS (prepolymer:curing agent = 10:1; Sylgard 184, Dow Corning) is spun (1,170 rpm) onto the lenses. The PDMS is cured for 2 h at 60°C.

10. A photoresist-coated wafer (500 nm thick; S1805 spun at 4,000 rpm) is prepared as described previously (*see* **Section 3.1.2**)

11. The microlenses are positioned at a distance of ∼40 cm above the surface of the overhead projector. The photoresist-coated wafer is placed in contact with the PDMS-coated microlenses.

12. The overhead projector (equipped with a standard broadband light source (λ = 400–1,200 nm) is turned on for 1 min to expose the photoresist.

13. The photoresist is developed in dilute developer (Microposit 351:water = 5:1)

3.2. Fabrication of Poly(dimethylsiloxane) Replicas

Microstructures made in photoresist are usually not directly useful for applications in biotechnology; they cannot be transferred from one substrate to another easily, be handled independently from the substrate, or be modified chemically. Replication of the topography of the photoresist into an elastomeric polymer overcomes these limitations. Poly(dimethylsiloxane) (PDMS) is used commonly as the elastomeric material for replicas in soft lithography because it has eight useful properties: (1) it is inexpensive; (2) it is commercially available; (3) it is chemically inert; (4) it is non-toxic; (5) it has a low surface free energy (21.6 dyn/cm^2) (25, 26); (6) it is optically transparent (25, 26); (7) it is intrinsically hydrophobic, but its surface can be made hydrophilic through brief treatment to an oxygen plasma; and (8) it is biocompatible (27, 30, 31).

Poly(dimethylsiloxane) replicas are prepared by molding of the liquid prepolymer against a photoresist master (**Fig. 3.6**). PDMS stamps (or molds) used for printing or microfluidics are thicker than the height of the features on the master. PDMS membranes are thinner than the height of the features on the

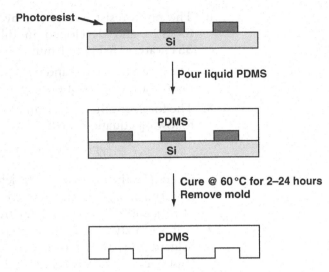

Fig. 3.6. Schematic diagram of replica molding of a master into PDMS.

master, so when released from the master these membranes have arrays of holes. Detailed procedures for the fabrication of each type of PDMS replica are described in the following subsections.

3.2.1. PDMS Stamps (or Molds)

Poly(dimethylsiloxane) stamps with thicknesses > 1 mm are prepared by pouring the liquid pre-polymer over the photoresist-based master. The thickness of the stamp does not have to be controlled precisely for applications in printing and in microfluidics.

1. Fabricate a "master" – a patterned photoresist on a solid substrate – by following the procedures described in **Section 3.2**.

2. Place the master in a vacuum dessicator along with a vial containing a few droplets of $CF_3(CF_2)_6(CH_2)_2SiCl_3$ under vacuum (~100 mTorr) for 20 min. The $CF_3(CF_2)_6(CH_2)_2SiCl_3$ will react with the exposed surface of the PDMS to modify the surface chemically, which decreases the surface free energy from 21.6 dyn/cm^2 to ~19 dyn/cm^2. Lowering the surface free energy of the master facilitates removal of the PDMS replica from the photoresist master without causing damage to the master or the replica.

3. Prepare a mixture of the PDMS prepolymer and the curing agent in a ratio of 10:1 by weight and mix thoroughly. Remove the gas bubbles that develop as a result of the mixing by placing the mixture in a dessicator under vacuum (~100 mTorr) for 45 min.

4. Pour the liquid PDMS over the master and cure to an elastomeric solid at 60°C for 3–4 h.

5. Use a scalpel or razor blade to cut around the features on the master and manually peel the PDMS stamp off the master (*see* **Note 9**).

3.2.2. PDMS Membranes

It is possible to prepare thin (<1 mm) membranes of PDMS by spin coating (40) and compression molding (31, 41). The thickness of the membranes is tailored to be less than the height of the features in the master so that an array of holes will exist in the membrane. These holes (when the membrane is placed in contact with a flat substrate) serve as microwells that are useful for biological experiments such as the patterning of cells on surfaces (**Fig. 3.7**) (40). Here we describe the procedure to prepare PDMS membranes by spin coating. This procedure is the same as that described in **Section 3.4.1**, up to Step 4. The subsequent steps are replaced with the following instructions:

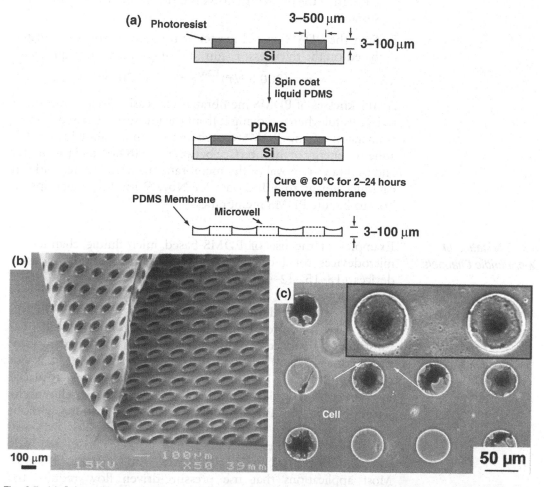

Fig. 3.7. (a) Schematic diagram of the fabrication of PDMS membranes by replica molding. (b) Scanning electron micrograph of a PDMS membrane with holes that was prepared by spin coating PDMS onto an array of photoresist posts. (c) Optical micrograph of cells patterned in the holes in a PDMS membrane. The surface of the PDMS was modified by physisorption of bovine serum albumin (BSA) to prevent the adhesion of cells to the membrane. The regions of the substrate exposed through the holes in the PMDS membrane were modified by physisorption of fibronectin to promote the adhesion of cells in these regions. The images in (b) and (c) are reprinted with permission from Ostuni et al. (40) © 2000 American Chemical Society.

1. Manually place the master onto the chuck of the spin coater.

For membranes > 30 μm thick use the following procedure:

2. Pour enough PDMS onto the master to coat the entire surface.

3. Spin the PDMS-coated master at the rate, r (rpm), calculated based on the thickness, τ, using the following equations:

$$r = (298790/\tau)^{0.8632}; \ \tau = 150-600 \ \mu m$$
$$r = (158599/\tau)^{0.9563}; \ \tau = 30-150 \ \mu m$$

For membranes <30 μm thick use the following procedure:

4. Dilute the PDMS with heptane at a 1:1 ratio by weight; pour enough of the diluted PDMS onto the master to coat the entire surface.

5. Spin the PDMS-coated master at the rate, r (rpm), calculated based on the thickness, τ (μm), using the following equation:

$$r = (200.59/\tau)^{3.267}; \ \tau = 15-30 \ \mu m$$

All thicknesses of PDMS membranes tear easily, so it is important to be careful when removing it from the photoresist-based master. Immersion of the PDMS-coated master in ethanol lowers the adhesive energy of the interface between the PDMS and the master and facilitates removal of the membrane from the master without causing damage to either one. *See* **Note 9** for additional tips on how to handle PDMS membranes.

3.3. Fabrication of Microfluidic Channels

Examples of the use of PDMS-based microfluidic channels in microdevices for biological applications include immunoassay devices (13–15, 42–46), DNA and protein separators (47–51), cell sorters (52–55), and tools for cell biology (56–59). PDMS-based microfluidic devices are prepared by sealing the PDMS replica of a photoresist master against a topographically flat substrate; the surface of this substrate forms the fourth wall of the channel (27, 60). PDMS will seal reversibly (i.e., it is not chemically bonded) by van der Waals interactions when placed in contact with a clean surface; the reversible seal allows the separation of the PDMS replica and the substrate without causing damage to either surface (60). Oxidized PDMS will seal permanently (i.e., it is covalently bonded) to clean oxide surfaces (e.g., glass, oxidized PDMS, and quartz) (27, 60). Most applications that use pressure-driven flow require the PDMS to be sealed permanently to prevent rupture of the seal between the PDMS and the flat substrate. **Figure 3.8** shows the preparation of a microfluidic channel by sealing a PDMS mold to glass (*see* **Note 10**). The procedure used to fabricate a 3-inlet, 1-outlet microfluidic network is outlined below:

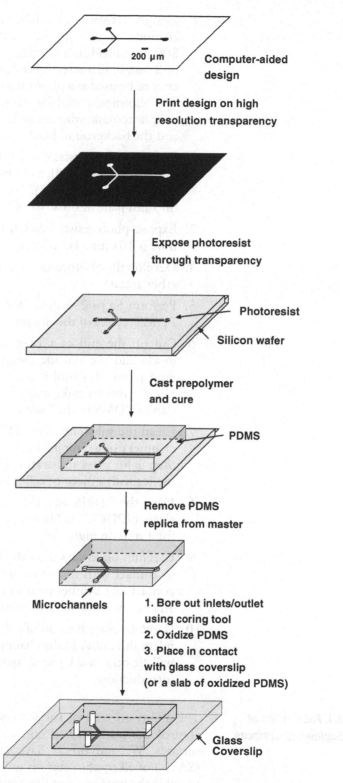

Computer-aided
design

Print design on high
resolution transparency

Expose photoresist
through transparency

Photoresist

Silicon wafer

Cast prepolymer
and cure

PDMS

Remove PDMS
replica from master

Microchannels

1. Bore out inlets/outlet
using coring tool
2. Oxidize PDMS
3. Place in contact
with glass coverslip
(or a slab of oxidized PDMS)

Glass
Coverslip

200 µm

Fig. 3.8. Schematic diagram of the process of fabrication of PDMS-based microfluidic channels by soft lithography.

1. Design networks of microfludic channels such that three channels of 100 μm in width intersect into a channel of 300 μm in width. The inlets and outlets are drawn as circles of 2 mm in diameter. This design is printed onto a transparency to be used as a photomask as described previously. Note: The photoresist used for this procedure is a negative resist, so the photomask, when printed, should have the channels clear and the background black.

2. Prepare a photoresist-coated substrate by spinning Microposit SU-8-2050 (Microchem Corp., Newton, MA) at 3,000 rpm (*see* **Note 2**). Bake the photoresist-coated wafer on a hot plate at 65°C for 3 min and at 95°C for 6 min.

3. Expose photoresist through the photomask. The exposure time is 40 s for a lamp intensity of $10 \text{ mJ cm}^{-2}\text{s}^{-1}$ at 365 nm.

4. Develop the photoresist for 6 min in propylene glycol methyl ether acetate.

5. Perform Steps 2–5 described in **Section 3.4** to prepare the PDMS replica of the master.

6. Cut off the end of a 16-gague syringe needle and file the inside and the outside using a pair of sharp tweezers and sand paper; this tool is used to form the inlets and outlets. Use the tool to make holes from the bottom to the top of the slab of PDMS in the location of the inlets and the outlets.

7. Clean the substrate that will form the bottom surface of the channel using a piranha solution described in **Section 2.3.2** Step 1 for glass substrates. Clean the surface of the PDMS replica with a piece of cellophane tape.

8. Place the PDMS and the substrate into an oxygen plasma cleaner (PDC-32G Harrick, or equivalent device) and oxidize for 1 min on high.

9. Carefully place the side of the PDMS with the channel in relief in contact with the substrate and allow the two surfaces to contact one another conformally. Place the sealed channel in an oven at 60°C for 1 h to improve the quality of the seal.

10. Carefully place the end of a PE-60 polyethylene tube into the inlets and outlet of the channel. The opposite end of the tube will fit onto a 21-gague syringe needle and permit sample introduction.

3.4. Fabrication of Engineered Surfaces

The ability to control surface chemistry is useful in biology to control the position and concentration of proteins or cells on surfaces. We and others have used self-assembled monolayers (SAMs) of alkanethiolates on gold, silver, copper, and palladium to tailor the properties of these surfaces, for example, to resist the adsorption of proteins or to present specific ligands for biospecific binding of cells or proteins (61–76). Patterns of SAMs on these types of surfaces are generated by microcontact printing (μCP) (**Fig. 3.9a**). In μCP, a PDMS stamp is wet with a molecular "ink." The ink is transferred in the regions of contact between the stamp and the surface. This technique can also pattern proteins (77–82) and DNA directly (83, 84). Below we describe the protocol for engineering surfaces: (1) to control the location and shape of cell cultures for use in cell biology experiments (64) and (2) to display biologically relevant ligands for use in immunoassays (85). We limit our discussion to the rapid prototyping of these types of surfaces and refer the reader to the literature for instructions on how to carry out such experiments (57, 64, 74, 85, 86).

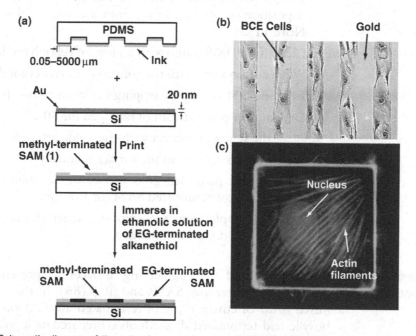

Fig. 3.9. (**a**) Schematic diagram of the procedure used to prepare SAMs for patterning cells by microcontact printing. (**b**) Optical micrograph of BCE cells grown on an Au substrate pattered with a SAM of methyl-terminated alkanethiols (regions where cells stick) and a SAM of EG-terminated alkanethiol (regions where the cells do not stick). (**c**) Fluorescence micrograph of a cell grown on a surface patterned by this technique. The actin and the nucleus were stained with fluorescently labeled molecules. These images are reprinted with permission from Ostuni et al. (64) © 2001 American Chemical Society.

3.4.1. Microcontact Printing of SAMs to Pattern Cells

Surfaces for cell cultures are prepared by patterning a methyl-terminated alkanethiol in the regions where adhesion of the cells is desired and an ethylene glycol (EG)-terminated alkanethiol in the regions where adhesion is not desired. A discussion of why ethylene glycol-terminated SAMs resist the non-specific adsorption of proteins is presented in detail elsewhere (87). **Figure 3.9b** and **c** show examples of the use of such a surface to pattern the location and shape of cells (64, 86). The following procedure describes this process:

1. Prepare a photoresist master and a PDMS stamp as described in **Sections 3.2** and **3.4.1**. The stamp should be designed such that the raised features represent regions where cells will adsorb, and the recessed regions are where cells will not adsorb.

2. Prepare an ethanolic solution of 1-octadecanethiol (10 mM).

3. Prepare an ethanolic solution of tri(ethylene glycol)-terminated alkanethiol (2 mM).

4. Prepare a gold-coated silicon wafer (or glass slide) (20 nm gold/2 nm titanium as an adhesion layer) by physical vapor deposition or by purchasing from a commercial source (*see* **Note 11**).

5. Clean the PDMS stamp with a piece of cellophane tape.

6. Wet a cotton swab with the solution of octadecanethiol.

7. Ink the PDMS stamp by swiping the swab across the surface.

8. Dry the stamp in a stream of nitrogen for 30 s.

9. Place the stamp in contact with the gold surface for 5–10 s.

10. Remove the stamp from the surface manually.

11. Immerse the patterned gold surface in the solution of the ethylene glycol-terminated SAM for 1–5 min.

12. Remove sample from the solution, wash thoroughly with ethanol, and dry under nitrogen.

3.4.2. Microcontact Printing of Reactive SAMs to Pattern Ligands

A useful method for engineering surfaces to present specific ligands is to use reactive SAMs and μCP (85). In this method, a mixed SAM of ethylene glycol-terminated alkanethiols and carboxylic acid-terminated alkanethiols is prepared on a gold surface. The carboxylic acid-terminated alkanethiol is activated chemically to allow reaction with amino groups at high yield. Ligands containing amino groups are patterned on the surface by μCP (**Fig. 3.10a**). These patterned surfaces are useful for biological experiments such as immunoassays (**Fig. 3.10b** and **c**) (85). The procedure to prepare a biotin-presenting surface by this technique is described in detail below:

1. Prepare an ethanolic solution containing both hexa(ethylene glycol)-terminated undecanethiol (2 mM) and tri(ethylene glycol)-terminated undecanethiol (2 mM).

2. Prepare a gold-coated silicon wafer (or glass slide) (20 nm gold/2 nm titanium as an adhesion layer) by physical vapor deposition or by purchasing from a commercial source (*see* **Note 11**).

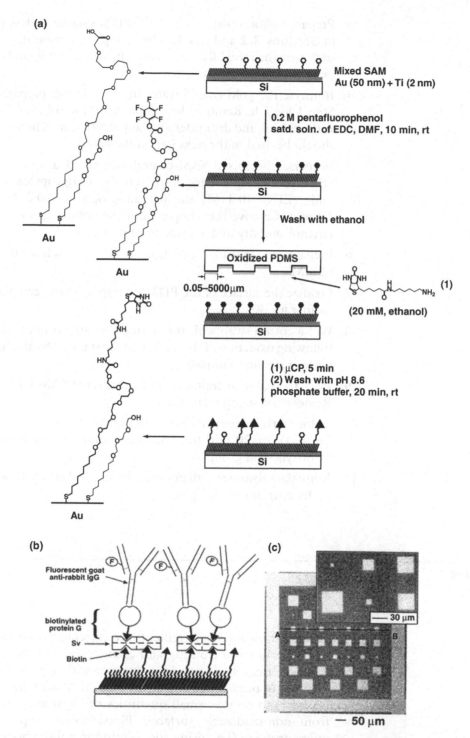

Fig. 3.10. (a) Schematic diagram for the patterning of reactive SAMs by microcontact printing. (b) Schematic diagram of an immunoassay performed on a sample patterned by this technique. (c) Fluorescence micrograph of patterned surfaces with fluorescently labeled antibodies bound to SAMs presenting biotin groups. The scheme in (b) and the fluorescence micrograph in (c) are reprinted permission from Lahiri et al. (85) © 1999 American Chemical Society.

3. Prepare a photoresist master and a PDMS stamp as described in **Sections 3.2** and **3.4.1**. The stamp should be designed such that the raised features represent where the ligands will be patterned on the surface.

4. Immerse the gold-coated wafer in the solution prepared in Step 1 for 2 h. Remove the sample from the solution, wash with ethanol, and dry under a stream of nitrogen. The samples should be used in the next step immediately.

5. Immerse the mixed-SAM-coated surface in a solution of DMF containing 1-ethyl-3-(dimethylamino)propylcarbodiimide (EDC) (0.1 M) and pentafluorophenol (0.2 M) for 10 min. Remove the sample from the solution, wash with ethanol, and dry under a stream of nitrogen.

6. Prepare an ethanolic solution of biotin cadaverine (1) (2 mM).

7. Oxidize the surface of the PDMS stamp in an oxygen plasma cleaner for 30 s.

8. Wet a cotton swab with the solution in Step 6 immediately following oxidation of the surface of the stamp. Dry the stamp under a stream of nitrogen.

9. Place the stamp in contact with the activated SAM for 5 min. Remove the stamp manually.

10. Immerse the patterned surface in phosphate buffer (pH 8.6, 25 mM) for 20 min to hydrolyze the remaining activated carboxylic acid groups.

11. Rinse the substrates with deionized water and ethanol and dry under a stream of nitrogen.

4. Notes

1. This solution and its ingredients are extremely corrosive. It will burn holes in clothing and cause severe skin burns on contact. It can also explode on contact with any significant quantity of organic solvent or material. It is safe when used (as here) to remove small quantities of organic impurities from non-oxidizable surfaces. Please follow appropriate safety protocol (i.e., using the solution in a fume hood and wearing thick rubber gloves, a lab coat, and goggles with side shields) while using this solution. Please read the Materials Safety Data Sheets (MSDS) for proper disposal instructions.

2. The files must be compatible with the plotters or printers used to make the transparencies. Check first with the commercial printing company to assure compatibility of your files with their printers.

3. The spin speeds, exposure conditions, and developing solutions vary for each type of photoresist. This information can be obtained from the companies that manufacture the resist. It is best to perform photolithography in a clean room facility to minimize contamination of the features by dust. Some experimentation is necessary to determine the proper exposure times; these times can vary widely from lamp to lamp.

4. Photolithography with the side without the ink in contact with the photoresist yields distorted features because of diffraction of light as it passes through the mask.

5. The mask and the surface of the substrate will be in focus at the same time only if the mask is in a conjugate image plane to that of the substrate. Most microscopes have a lens that expands the image to fill the back aperture of the objective. In a Leica DMRX microscope, this objective reduces the reduction power of the objective by a factor of 4.

6. The UV light in the room where this procedure is carried out must be minimized to prevent accidental exposure of the photoresist. The photoresist will not be exposed by short periods (<30 min) of illumination from standard fluorescent lighting.

7. This technique works with oil and water immersion lenses. A glass cover slip may be used to protect the surface of the resist from the oil, but is not necessary when using water.

8. Distortions in the size and shape of the photoresist features can occur at the edges of the lens array (>1 cm from the center of the array). It is therefore difficult to produce arrays of features by this technique that are uniform in shape and size over areas larger than 4 cm^2.

9. PDMS membranes can be difficult to handle because they are flexible and tear easily. It is useful to cure a thick (~1 mm) "ring" of PDMS outside of the area of the photoresist features. This ring of PDMS provides structural rigidity to the membrane, so the membranes can be handled easily using tweezers.

10. The sealing of PDMS to a slab of PDMS uses the same procedure as that of sealing PDMS to glass; it, however, requires plasma oxidation of *both* pieces of PDMS before they are placed in contact.

11. Gold substrates can be purchased from vendors such as Platypus Technologies (Madison, WI).

Acknowledgements

The authors thank the National Institutes of Health (GM065364) and DARPA for financial support.

References

1. Whitesides, G. M., Ostuni, E. S., Takayama, S., Jiang, X., and Ingber, D. E. (2001) Soft Lithography in Biology and Biochemistry. *Annu. Rev. Biomed. Eng.* 3, 335–373.

2. Henkel, G. W. (2002) Live cell assays: tools for functional genomics. *Funct. Genomics Ser.* 1, 279–311.

3. Bhadriraju, K., and Chen, C. S. (2002) Engineering cellular microenvironments to improve cell-based drug testing. *Drug Discov. Today* 7, 612–620.

4. Sundberg, S. A. (2000) High-throughput and ultra-high-throughput screening: solution- and cell-based approaches. *Curr. Opin. Biotechnol.* 11, 47–53.

5. Mere, L., Bennett, T., Coassin, P., England, P., Hamman, B., Rink, T., Zimmerman, S., and Negulescu, P. (1999) Miniaturized FRET assays and microfluidics: key components for ultra-high-throughput screening. *Drug Discov. Today* 4, 363–369.

6. Paegel Brian, M., Blazej Robert, G., and Mathies Richard, A. (2003) Microfluidic devices for DNA sequencing: sample preparation and electrophoretic analysis. *Curr. Opin. Biotechnol.* 14, 42–50.

7. Righetti, P. G., Gelfi, C., and D'Acunto, M. R. (2002) Recent progress in DNA analysis by capillary electrophoresis. *Electrophoresis* 23, 1361–1374.

8. Foquet, M., Korlach, J., Zipfel, W., Webb, W. W., and Craighead, H. G. (2002) DNA fragment sizing by single molecule detection in submicrometer-sized closed fluidic channels. *Anal. Chem.* 74, 1415–1422.

9. Cabodi, M., Turner, S. W. P., and Craighead, H. G. (2002) Entropic Recoil Separation of Long DNA Molecules. *Anal. Chem.* 74, 5169–5174.

10. Fodor, S. P. A. (1997) DNA sequencing: Massively parallel genomics. *Science* 277, 393, 395.

11. Kennedy, G. C., Matsuzaki, H., Dong, S., Liu, W.-M., Huang, J., Liu, G., Su, X., Cao, M., Chen, W., Zhang, J., Liu, W., Yang, G., Di, X., Ryder, T., He, Z., Surti, U., Phillips, M. S., Boyce-Jacino, M. T., Fodor, S. P. A., and Jones, K. W. (2003) Large-scale genotyping of complex DNA. *Nature Biotechnol.* 21, 1233–1237.

12. Lipshutz, R. J., Fodor, S. P. A., Gingeras, T. R., and Lockhart, D. J. (1999) High density synthetic oligonucleotide arrays. *Nature Genet.* 21, 20–24.

13. Sia, S. K., Linder, V., Parviz, B. A., Siegel, A., and Whitesides, G. M. (2004) An integrated approach to a portable and low-cost immunoassy for resource-poor settings. *Angew. Chem. Int. Ed.* 43, 498–502.

14. Linder, V., Verpoorte, E., De Rooij, N. F., Sigrist, H., and Thormann, W. (2002) Application of surface biopassivated disposable poly(dimethylsiloxane)/glass chips to a heterogeneous competitive human serum immunoglobulin G immunoassay with incorporated internal standard. *Electrophoresis* 23, 740–749.

15. Rossier, J. S., and Girault, H. H. (2001) Enzyme linked immunosorbent assay on a microchip with electrochemical detection. *Lab on a Chip* 1, 153–157.

16. Lai, S., Wang, S., Luo, J., Lee, L. J., Yang, S.-T., and Madou, M. J. (2004) Design of a compact disk-like microfluidic platform for enzyme-linked immunosorbent assay. *Anal. Chem.* 76, 1832–1837.

17. Song, J. M., Griffin, G. D., and Vo-Dinh, T. (2003) Application of an integrated microchip system with capillary array electrophoresis to optimization of enzymatic reactions. *Anal. Chim. Acta* 487, 75–82.

18. Stenger, D. A., Andreadis, J. D., Vora, G. J., and Pancrazio, J. J. (2002) Potential applications of DNA microarrays in biodefense-related diagnostics. *Curr. Opin. Biotechnol.* 13, 208–212.

19. Sadik, O. A., Land, W. H., Jr., and Wang, J. (2003) Targeting chemical and biological warfare agents at the molecular level. *Electroanalysis* 15, 1149–1159.

20. Reynolds, J. G., and Hart, B. R. (2004) Nanomaterials and their application to defense and homeland security. *JOM* 56, 36–39.

21. Needham, S. R. (2003) Bioanalytical method validation: example, HPLC/MS/MS bioanalysis of an anti-nerve gas agent drug for homeland security. *Am. Pharm. Rev.* 6, 86, 88, 90–91.

22. Yadav, P., and Blaine, L. (2004) Microbiological threats to homeland security. *IEEE Eng. Med. Bio. Mag.* 23, 136–141.

23. Kun, L. (2004) Technology and policy review for homeland security. *IEEE Eng. Med. Bio. Mag.* 23, 30–44.

24. Gluodenis, T., and Harrison, S. (2004) Homeland security and bioterrorism applications: detection of bioweapon pathogens by microfluidic-based electrophoretic DNA analysis. *Med. Lab. Observ.* 36, 34–38.

25. Xia, Y., and Whitesides, G. M. (1998) Soft Lithography. *Angew. Chem. Int. Ed.* 37, 550–575.

26. Xia, Y., and Whitesides, G. M. (1998) Soft Lithography. *Ann. Rev. Mater. Sci.* 28, 153–184.

27. Sia, S. K., and Whitesides, G. M. (2003) Microfluidic devices fabricated in poly(dimethylsiloxane) for biological studies. *Electrophoresis* 24, 3563–3576.

28. Whitesides, G. M., and Stroock, A. D. (2001) Flexible methods for microfluidics. *Phys. Today* 54, 42–48.

29. Schmid, H., and Michel, B. (2000) Siloxane polymers for high-resolution, high-accuracy soft lithography. *Macromolecules* 33, 3042–3049.

30. Jiang, X., and Whitesides, G. M. (2003) Engineering microtools in polymers to study cell biology. *Eng. Life Sci.* 3, 475–480.

31. Folch, A., and Toner, M. (2000) Microengineering of cellular interactions. *Annu. Rev. Biomed. Eng* 2, 227–256.

32. Linder, V., Wu, H., Jiang, X., and Whitesides, G. M. (2003) Rapid prototyping of 2D structures with feature sizes larger than 8 μm. *Anal. Chem.* 75, 2522–2527.

33. Qin, D., Xia, Y., and Whitesides, G. M. (1996) Rapid prototyping of complex structures with feature sizes larger than 20 μm. *Adv. Mater.* 8, 917–919.

34. Love, J. C., Wolfe, D. B., Jacobs, H. O., and Whitesides, G. M. (2001) Microscope projection photolithography for rapid prototyping of masters with micron-scale features for use in soft lithography. *Langmuir* 17, 6005–6012.

35. Wu, H., Odom, T. W., and Whitesides, G. M. (2002) Connectivity of features in microlens array reduction photolithography: Generation of different patterns using

36. Wu, H., Odom, T. W., and Whitesides, G. M. (2002) Reduction photolithography using microlens arrays: Applications in grayscale photolithography. *Anal. Chem.* 74, 3267–3273.

37. Wu, H., Odom, T. W., and Whitesides, G. M. (2002) Generation of chrome masks with micrometer-scale features using microlens lithography. *Adv. Mater.* 14, 1213–1216.

38. Wu, M. H., and Whitesides, G. M. (2001) Fabrication of arrays of two-dimensional micropatterns using microspheres as microlenses for projection lithography. *Appl. Phys. Lett.* 78, 2273–2275.

39. Wu, M.-H., Paul, K. E., Yang, J., and Whitesides, G. M. (2002) Fabrication of frequency-selective surfaces using microlens photolithography. *Appl. Phys. Lett.* 80, 3500–3502.

40. Ostuni, E., Kane, R., Chen, C. S., Ingber, D. E., and Whitesides, G. M. (2000) Patterning mammalian cells using elastomeric membranes. *Langmuir* 16, 7811–7819.

41. Folch, A., Jo, B.-H., Beebe, D., and Toner, M. (2000) Microfabricated elastomeric stensils for micropatterning cell cultures. *J. Biomed. Mater. Res.* 52, 346–353.

42. Hatch, A., Kamholz, A. E., Hawkins, K. R., Munson, M. S., Schilling, E. A., Weigl, B. H., and Yager, P. (2001) A rapid diffusion immunoassay in a T-sensor. *Nature Biotechnol.* 19, 461–465.

43. Chiem, N. H., and Harrison, D. J. (1998) Microchip systems for immunoassay: an integrated immunoreactor with electrophoretic separation for serum theophylline determination. *Clin. Chem.* 44, 591–598.

44. Ismagilov, R. F., Ng, J. M. K., Kenis, P. J. A., and Whitesides, G. M. (2001) Microfluidic arrays of fluid-fluid diffusional contacts as detection elements and combinatorial tools. *Anal. Chem.* 73, 5207–5213.

45. Bernard, A., Michel, B., and Delamarche, E. (2001) Micromosaic immunoassays. *Anal. Chem.* 73, 8–12.

46. Jiang, X., Ng, J. M. K., Stroock, A. D., Dertinger, S. K. W., and Whitesides, G. M. (2003) A miniaturized, parallel, serially diluted immunoassay for analyzing multiple antigens. *J. Am. Chem. Soc.* 125, 5294–5295.

47. Harris, C. M. (2003) Shrinking the LC landscape. *Anal. Chem.* 75, 64A–69A.

48. Slentz, B. E., Penner, N. A., Lugowska, E., and Regnier, F. (2001) Nanoliter capillary

electrochromatography columns based on collocated monolithic support structures molded in poly(dimethyl siloxane). *Electrophoresis* 22, 3736–3743.

49. Jiang, Y., Wang, P.-C., Locascio, L. E., and Lee, C. S. (2001) Integrated plastic microfluidic devices with ESI-MS for drug screening and residue analysis. *Anal. Chem.* 73, 2048–2053.

50. Kim, J. S., and Knapp, D. R. (2001) Microfabrication of polydimethylsiloxane electrospray ionization emitters. *J. Chromatogr. A* 924, 137–145.

51. Huikko, K., Oestman, P., Grigoras, K., Tuomikoski, S., Tiainen, V. M., Soininen, A., Puolanne, K., Manz, A., Franssila, S., Kostiainen, R., and Kotiaho, T. (2003) Poly(dimethylsiloxane) electrospray devices fabricated with diamond-like carbon-poly(dimethylsiloxane) coated SU-8 masters. *Lab on a Chip* 3, 67–72.

52. Fu, A. Y., Chou, H.-P., Spence, C., Arnold, F. H., and Quake, S. R. (2002) An integrated microfabricated cell sorter. *Anal. Chem.* 74, 2451–2457.

53. Deng, T., Prentiss, M., and Whitesides, G. M. (2002) Fabrication of magnetic microfiltration systems using soft lithography. *Appl. Phys. Lett.* 80, 461–463.

54. Cho, B. S., Schuster, T. G., Zhu, X., Chang, D., Smith, G. D., and Takayama, S. (2003) Passively driven integrated microfluidic system for separation of motile sperm. *Anal. Chem.* 75, 1671–1675.

55. Beebe, D., Wheeler, M., Zeringue, H., Walters, E., and Raty, S. (2002) Microfluidic technology for assisted reproduction. *Theriogenology* 57, 125–135.

56. Takayama, S., McDonald, J. C., Ostuni, E., Liang, M. N., Kenis, P. J. A., Ismagilov, R. F., and Whitesides, G. M. (1999) Patterning cells and their environments using multiple laminar fluid flows in capillary networks. *Proc. Natl. Acad. Sci. U.S.A.* 96, 5545–5548.

57. Chiu, D. T., Li Jeon, N., Huang, S., Kane, R. S., Wargo, C. J., Choi, I. S., Ingber, D. E., and Whitesides, G. M. (2000) Patterned deposition of cells and proteins onto surfaces by using three-dimensional microfluidic systems. *Proc. Natl. Acad. Sci. U.S.A.* 97, 2408–2413.

58. Takayama, S., Ostuni, E., LeDuc, P., Naruse, K., Ingber, D. E., and Whitesides, G. M. (2001) Subcellular positioning of small molecules. *Nature* 411, 1016.

59. Sawano, A., Takayama, S., Matsuda, M., and Miyawaki, A. (2002) Lateral propagation of EGF signaling after local stimulation is dependent on receptor density. *Dev. Cell* 3, 245–257.

60. McDonald, J. C., Duffy, D. C., Anderson, J. R., Chiu, D. T., Wu, H., and Whitesides, G. M. (2000) Fabrication of microfluidic systems in poly(dimethylsiloxane). *Electrophoresis* 21, 27–40.

61. Kane, R. S., Deschatelets, P., and Whitesides, G. M. (2003) Kosmotropes form the basis of protein-resistant surfaces. *Langmuir* 19, 2388–2391.

62. Herrwerth, S., Rosendahl, T., Feng, C., Fick, J., Eck, W., Himmelhaus, M., Dahint, R., and Grunze, M. (2003) Covalent coupling of antibodies to self-assembled monolayers of carboxy-functionalized poly(ethylene glycol): Protein resistance and specific binding of biomolecules. *Langmuir* 19, 1880–1887.

63. Veiseh, M., Zareie, M. H., and Zhang, M. (2002) Highly selective protein patterning on gold-silicon substrates for biosensor applications. *Langmuir* 18, 6671–6678.

64. Ostuni, E., Chapman, R. G., Liang, M. N., Meluleni, G., Pier, G., Ingber, D. E., and Whitesides, G. M. (2001) Self-assembled monolayers that resist the adsorption of proteins and the adhesion of bacterial and mammalian cells. *Langmuir* 17, 6336–6343.

65. Schwendel, D., Dahint, R., Herrwerth, S., Schloerholz, M., Eck, W., and Grunze, M. (2001) Temperature dependence of the protein resistance of poly- and oligo(ethylene glycol)-terminated alkanethiolate monolayers. *Langmuir* 17, 5717–5720.

66. Ostuni, E., Chapman, R. G., Holmlin, R. E., Takayama, S., and Whitesides, G. M. (2001) A survey of structure-property relationships of surfaces that resist the adsorption of protein. *Langmuir* 17, 5605–5620.

67. Zhu, B., Eurell, T., Gunawan, R., and Leckband, D. (2001) Chain-length dependence of the protein and cell resistance of oligo(ethylene glycol)-terminated self-assembled monolayers on gold. *J. Biomed. Mater. Res.* 56, 406–416.

68. Holmlin, R. E., Chen, X., Chapman, R. G., Takayama, S., and Whitesides, G. M. (2001) Zwitterionic SAMs that resist nonspecific adsorption of protein from aqueous buffer. *Langmuir* 17, 2841–2850.

69. Chapman, R. G., Ostuni, E., Liang, M. N., Meluleni, G., Kim, E., Yan, L., Pier, G., Warren, H. S., and Whitesides, G. M. (2001) Polymeric thin films that resist the adsorption of proteins and the adhesion of bacteria. *Langmuir* 17, 1225–1233.

70. Chapman, R. G., Ostuni, E., Takayama, S., Holmlin, R. E., Yan, L., and Whitesides, G. M. (2000) Surveying for surfaces that resist the adsorption of proteins. *J. Am. Chem. Soc.* 122, 8303–8304.

71. Chapman, R. G., Ostuni, E., Yan, L., and Whitesides, G. M. (2000) Preparation of mixed Self-Assembled Monolayers (SAMs) that resist adsorption of proteins using the reaction of amines with a SAM that presents interchain carboxylic anhydride groups. *Langmuir* 16, 6927–6936.

72. Ostuni, E., Yan, L., and Whitesides, G. M. (1999) The interaction of proteins and cells with self-assembled monolayers of alkanethiolates on gold and silver. *Colloids Surf.* 15, 3–30.

73. Pertsin, A. J., Grunze, M., and Garbuzova, I. A. (1998) Low-energy configurations of methoxy triethylene glycol terminated alkanethiol self-assembled monolayers and their relevance to protein adsorption. *J. Phys. Chem. B* 102, 4918–4926.

74. Mrksich, M., Dike, L. E., Tien, J., Ingber, D. E., and Whitesides, G. M. (1997) Using microcontact printing to pattern the attachment of mammalian cells to self-assembled monolayers of alkanethiolates on transparent films of gold and silver. *Exper. Cell Res.* 235, 305–313.

75. Mrksich, M., Chen, C. S., Xia, Y., Dike, L. E., Ingber, D. E., and Whitesides, G. M. (1996) Controlling cell attachment on contoured surfaces with self-assembled monolayers of alkanethiolates on gold. *Proc. Nat. Acad. Sci. U.S.A.* 93, 10775–10778.

76. DiMilla, P. A., Folkers, J. P., Biebuyck, H. A., Haerter, R., Lopez, G. P., and Whitesides, G. M. (1994) Wetting and Protein Adsorption on Self-Assembled Monolayers of Alkanethiolates Supported on Transparent Films of Gold. *J. Am. Chem. Soc.* 116, 2225–2226.

77. Donzel, C., Geissler, M., Bernard, A., Wolf, H., Michel, B., Hilborn, J., and Delamarche, E. (2001) Hydrophilic poly(dimethylsiloxane) stamps for microcontact printing. *Adv. Mater.* 13, 1164–1167.

78. Bernard, A., Renault, J. P., Michel, B., Bosshard, H. R., and Delamarche, E. (2000) Microcontact printing of proteins. *Adv. Mater.* 12, 1067–1070.

79. Graber, D. J., Zieziulewics, T. J., Lawrence, D. A., Shain, W., and Turner, J. N. (2003) Antigen binding specificity of antibodies patterned by microcontact printing. *Langmuir* 19, 5431–5434.

80. Renault, J. P., Bernard, A., Bietsch, A., Michel, B., Bosshard, H. R., Delamarche, E., Kreiter, M., Hecht, B., and Wild, U. P. (2003) Fabricating arrays of single protein molecules on glass using microcontact printing. *J. Phys. Chem. B* 107, 703–711.

81. Renault, J. P., Bernard, A., Juncker, D., Michel, B., Bosshard, H. R., and Delamarche, E. (2002) Fabricating microarrays of functional proteins using affinity contact printing. *Angew. Chem. Int. Ed.* 41, 2320–2323.

82. Tan, J. L., Tien, J., and Chen, C. S. (2002) Microcontact printing of proteins on mixed self-assembled monolayers. *Langmuir* 18, 519–523.

83. Xu, C., Taylor, P., Ersoz, M., Fletcher, P. D. I., and Paunov, V. N. (2003) Microcontact printing of DNA-surfactant arrays on solid substrates. *J. Mater. Chem.* 13, 3044–3048.

84. Lange, S. A., Benes, V., Kern, D. P., Hoerber, J. K. H., and Bernard, A. (2004) Microcontact printing of DNA molecules. *Anal. Chem.* 76, 1641–1647.

85. Lahiri, J., Ostuni, E., and Whitesides, G. M. (1999) Patterning ligands on reactive SAMs by microcontact printing. *Langmuir* 15, 2055–2060.

86. Chen, C. S., Mrksich, M., Huang, S., Whitesides, G. M., and Ingber, D. E. (1997) Geometric control of cell life and death. *Science* 276, 1425-1428.

87. Ostuni, E., Grzybowski, B., Mrksich, M., Roberts, C. S., and Whitesides, G. M. (2003) Adsorption of proteins to hydrophobic sites on mixed self-assembled monolayers. *Langmuir* 19, 1861–1872.

Chapter 4

Chemical Synthesis in Microreactors

Paul Watts and Stephen J. Haswell

Abstract

To develop a new generation of drugs, pharmaceutical companies need to be able to synthesise and screen novel chemicals with enhanced speed. New technology that would enable a cost neutral step change in the number of potential drug candidates would provide a distinct competitive advantage. Indeed the miniaturisation of chemical reactors offers many fundamental and practical advantages of relevance to the pharmaceutical industry, who are constantly searching for controllable, information-rich, high-throughput, environmentally friendly methods of producing products with a high degree of chemical selectivity. This chapter reviews the current and future applications of microreactors that could enhance the drug discovery process.

Key words: Chemical synthesis, microreactor, separation, electroosmotic flow, EOF, peptide synthesis, drug discovery, pharmaceutical.

1. Introduction

The success of pharmaceutical companies largely depends on their ability to synthesise novel chemical entities and to optimise the production of marketable drugs. In an industry where development costs are extraordinarily high, it is important to develop new technology that enables the rapid synthesis and screening of novel chemical entities. The miniaturisation of chemical reactors offer many advantages of relevance to the pharmaceutical industry, who are constantly searching for high-throughput methods of producing products with a high degree of chemical selectivity 114(1–8). This chapter illustrates how miniaturisation may revolutionise medicinal chemistry and the pharmaceutical industry through the implementation of innovative technology.

M.P. Hughes, K.F. Hoettges (eds.), *Microengineering in Biotechnology*, Methods in Molecular Biology 583, DOI 10.1007/978-1-60327-106-6_4, © Humana Press, a part of Springer Science+Business Media, LLC 2010

In their simplest form, microreactors consist of a network of micron-sized channels (10–300 μm in diameter) etched into a solid substrate. A number of materials such as silicon, quartz, glass, metals and polymers have been used to construct microreactors (9). Important considerations in material choice include chemical compatibility, ease, and reproducibility of fabrication, and whether the material supports electroosmotic flow (EOF) (10–13) with the solvents of interest and compatibility with detection methods. Glass is a popular choice for chemical synthesis since it allows EOF with many common solvents, is chemically inert, enables the use of a range of visible light detection methods and fabrication techniques are well established. For glass microreactors, which are ideal for synthetic chemistry, photolithographic fabrication of the channel network is performed (14, 15).

For solution-based chemistry, the channel networks are connected to a series of reservoirs containing chemical reagents to form the complete device (or "lab-on-a-chip") with overall dimensions of a few centimetre, as shown in **Fig. 4.1**. In the microreactor reagents can be brought together in a specific sequence, mixed and allowed to react for a specified time in a controlled region of the channel network using electrokinetic or hydrodynamic pumping. For electrokinetically driven systems, electrodes are located in the appropriate reservoirs (as illustrated in **Fig. 4.1**) to which specific voltage sequences can be delivered under automated computer control. This technique offers a simple but effective method of moving and separating reactants and products within a microreactor, without the need for moving parts. In comparison, hydrodynamic pumping exploits conventional or microscale pumps, notably syringe-type pumps to manoeuvre solutions around the channel network. However, hydrodynamic pumping has the disadvantage of requiring either an interface between large external pumps and the chip or the need to fabricate complex small moving parts within the device itself.

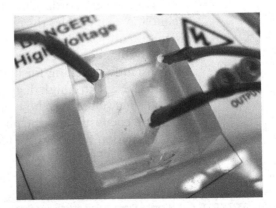

Fig. 4.1. Reaction in process within a microreactor under EOF control.

Until recently, the greatest research effort in the miniaturisation field has been in the area of analytical science, where the aim was to develop miniaturised total analytical systems (µTAS) (16–19). Alongside the continuing development of this area, a concerted effort has also been underway to establish the benefits that microreactors can bring to the field of reaction chemistry. The ability to manipulate reagent concentrations in both space and time by electrokinetic voltage control within the channel network of a microreactor (20–24) provides a level of reaction control that is not attainable in bulk stirred reactors where concentrations are generally uniform. Many chemical reactions have now been demonstrated to show improved reactivity, product yield and selectivity when performed in microreactors compared to those generated using conventional laboratory practices (25).

The outcome of the reported research has confirmed that microreactor methodology is applicable to performing both gas- and liquid-phase reaction chemistry (25). The evidence is that the unique modus operandi of microreactors, namely the low-volume spatial and temporal control of reactants and products in a diffusive mixing environment, in which distinctive thermal and concentration gradients exist, offers a novel method for chemical manipulation and generation of products. Reactions performed in a microreactor invariably generate comparatively pure products in high yield, when compared to the equivalent bulk reactions, in much shorter times (25). One of the immediate and obvious applications is therefore in drug and process discovery, where the generation of compounds either with different reagents or under variable conditions is an essential factor.

Performing chemical reactions within a microfluidic system also provides the opportunity to perform real-time separations. Hence, integration of a microreactor device, via a purification step, with one of the many highly sensitive microchannel-based biological assay systems would provide a drug discovery tool. This level of integrated functionality within one device clearly addresses some of industries' requirements for rapid compound production and screening. Apart from the greatly reduced reaction times demonstrated for the microreactors, handling times to assay and chemical reagent costs are virtually eliminated with this proposed technology. This is shown diagrammatically in **Fig. 4.2**.

This chapter describes how chemical synthesis and purification may be performed within microfluidic devices, which if integrated to a bioassay system would fulfil the above criteria. In particular, this chapter concentrates on the preparation of peptide molecules, which are present in many pharmaceutically important compounds and consequently the pharmaceutical industry is particularly interested in developing improved methodology for their preparation.

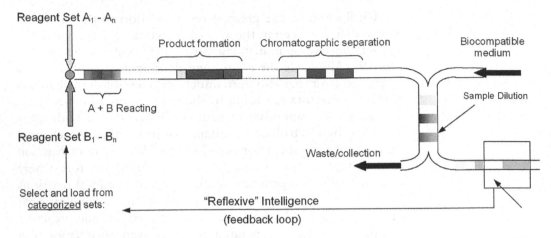

Fig. 4.2. Integration of a microreactor to a biological assay system for drug discovery.

The methods described below outline (1) how a microreactor may be operated by electroosmotic flow, (2) how peptide synthesis may be conducted within a microreactor and (3) how the technique may be extended to purify the product in situ within the microreactor.

2. Electroosmotic Flow

To illustrate the principles of EOF, one can consider a microchannel fabricated from a material having negatively charged functional groups on the surface. If a liquid, displaying some degree of dissociation, is brought into contact with the surface, positive counter ions will form a double layer such that the positively charged ions are attracted to the negatively charged surface. Application of an electric field causes this layer to move towards the negative electrode, thus causing the bulk liquid to move within the channel (**Fig. 4.3**). Electroosmotic flow is not just restricted to aqueous systems and a range of organic solvents can be mobilised using the technique (*see* **Note 1**). A major advantage of EOF is that it gives a high degree of spatial and temporal control to reactants mixing under a diffusive regime. Electroosmotic flow also has the advantage that the direction and magnitude of the flow may be readily changed by altering the applied voltage, where the flow rates increase with voltage. The flow rates, which are in the order of $\mu l \ min^{-1}$, are reproducible using EOF. In addition, the pumping technique has no mechanical or moving parts; hence, it is highly suitable for operation in miniaturised systems. An automated

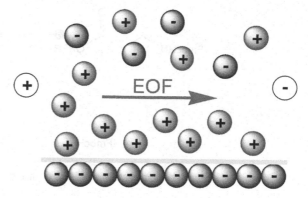

Fig. 4.3. Principle of electroosmotic flow.

power supply with multiple electrodes can support several pumping channels (*see* **Note 2**). Importantly, when using EOF, plugs of fluid are transported without significant hydrodynamic dispersion.

In order to perform EOF within a device

1. Prime the microreactor by filling the channels with the reaction solvent.

2. Using a syringe fitted with a pipette tip, push the solvent through the microreactor to ensure that there are no bubbles in the channel, as this will prevent EOF.

If EOF is absent or very slow it may be improved by the following:

1. Washing the channels of the microreactor with 0.1 M nitric acid for about 10 min in order to remove any contaminants from the channel.

2. Washing the channels with 0.1 M potassium hydroxide in order to regenerate the SiO⁻ surface

3. Peptide Synthesis in a Microreactor (26–28)

To conduct a chemical reaction within the microreactor, a solution of the pentafluorophenyl ester of Fmoc-β-alanine **1** (50 µl, 0.1 M; Fluka; Sigma-Aldrich Co.) in anhydrous DMF (Fluka; Sigma-Aldrich Co.) was added to reservoir A and a solution of amine **2** (50 µl, 0.1 M) was placed in reservoir B (**Fig. 4.4**). Anhydrous DMF (40 µl) was placed in reservoir C, which was used to collect the products of the reaction. It was found that using continuous flow of both reagents, where the ester **1** was maintained at 700 V and the amine **2** was maintained at 600 V, dipeptide **3** was produced quantitatively in 20 min. Product conversions were based on the amount of pentafluorophenyl ester **1** remaining in the

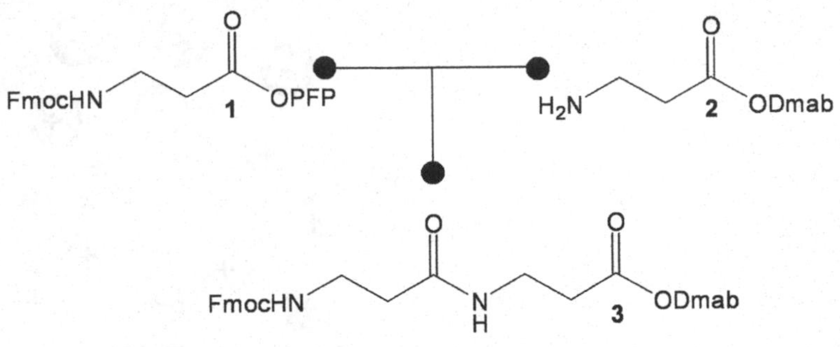

Fig. 4.4. Microreactor manifold for peptide synthesis.

sample. This represented a significant increase in yield compared with solution phase synthesis, where lengthy reaction times are also required (26–28).

The experiment was conducted at room temperature for a period of 20 min (the contents of reservoir C were analysed by high performance liquid chromatography, Jupiter C_{18}10 μm, 4.6 × 250 mm, with mobile phase composition 0.1% TFA (Sigma Aldrich Co.) in water and 0.1% TFA in acetonitrile (Fisher Scientific UK), using a gradient system of 30% aqueous to 70% aqueous in 20 min with a flow rate of 2.5 ml min^{-1} at room temperature).

Having demonstrated that peptide bonds could be successfully formed within a microreactor, the methodology was then extended to the preparation of longer chain peptides. Using the microreactor, the Dmab ester of Fmoc-β-alanine 4 (50 μl, 0.1 M) was placed in reservoir A and reacted with one equivalent of 1,8-diazabicyclo[5.4.0]undec-7-ene (DBU, Sigma Aldrich Co.) (50 μl, 0.1 M) in reservoir B, to give the free amine 2 in quantitative conversion. The authors then reacted the amine in situ with the pentafluorophenyl ester 5 (50 μl, 0.1 M) in reservoir C to give the dipeptide 6 (Fig. 4.5). Using continuous flow of the reagents the Dmab ester 4 was maintained at 750 V, DBU at 800 V and PFP ester 5 at 700 V; 96% overall conversion was obtained.

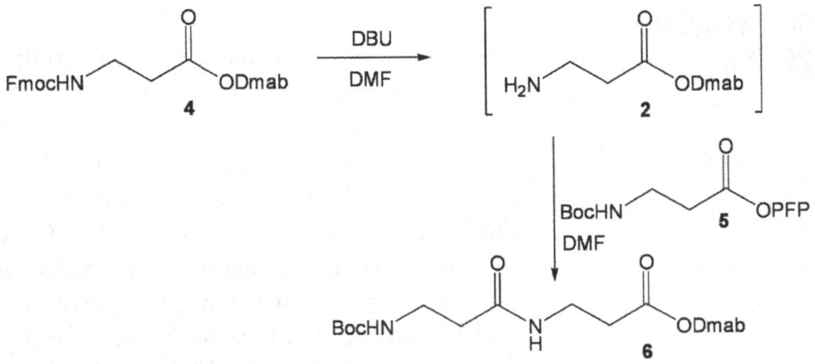

Fig. 4.5. Reaction of amine with pentafluorophenyl ester 5 to produce a dipeptide.

Having shown that more complex peptides could be produced by removal of the *N*-protecting group (namely Fmoc), the methodology was extended to the removal the Dmab protecting group using hydrazine (Sigma Aldrich Co.). Hence, a solution of the Dmab ester of Fmoc-β-alanine **4** (50 μl, 0.1 M) in anhydrous DMF was added to reservoir A, a solution of hydrazine (50 μl, 0.1 M) was placed in reservoir B and anhydrous DMF (40 μl) was placed in reservoir C. Using continuous flow of both reagents maintained at 700 V, quantitative deprotection was observed to give carboxylic acid **7** (**Fig. 4.6**). Significantly, deprotection was achieved in the micro reactions using just one equivalent of hydrazine, compared with 2% solutions used in solid phase peptide synthesis. This suggests that reactions conducted within microreactors are highly atom efficient.

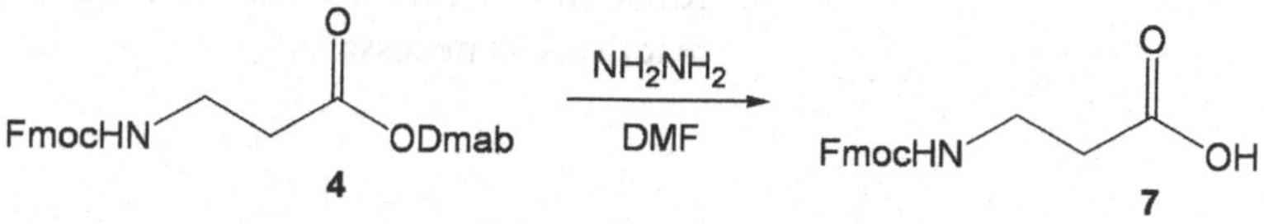

Fig. 4.6. Using solutions of the Dmab ester of Fmoc-b-alanine and of hydrazine, in anhydrous DMF under continuous flow conditions maintained at 700 V, quantitative deprotection was observed to give carboxylic acid.

The authors have further extended the approach to the synthesis of tripeptide **9** using exactly the same methodology. Reaction of pentafluorophenyl ester **1** with amine **2** formed dipeptide **3**, which was reacted with DBU to effect Fmoc deprotection. The amine **8** was then reacted in situ with another equivalent of pentafluorophenyl ester **1** to prepare tripeptide **9** in 30% overall conversion (**Fig. 4.7**). This approach clearly demonstrates that intermediates may be generated in situ and used in subsequent reactions enabling the combinatorial synthesis of peptides, which are of biological and pharmaceutical interest.

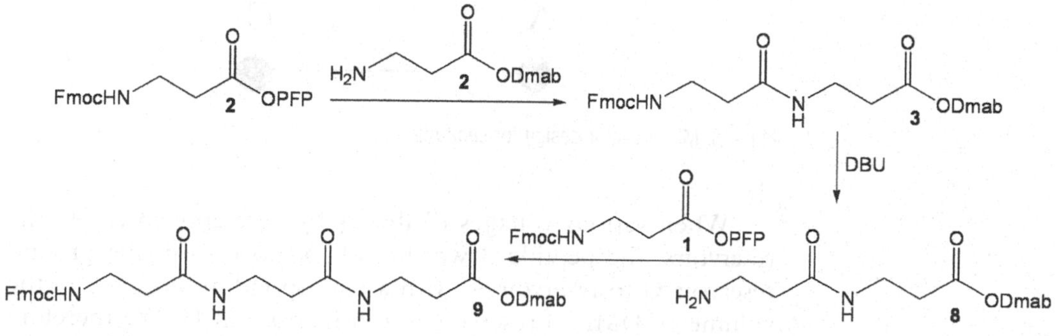

Fig. 4.7. The reaction of an amine with an equivalent of pentafluorophenyl ester to prepare tripeptide in 30% overall conversion.

General procedure when performing chemical synthesis in microreactors is as follows:

1. Ensure that there are no bubbles in the reactor channels.

2. Place known volumes of reagents and/or solvents (typically 50 μl) in the reservoirs using a microlitre syringe.

3. After conducting the reaction for a designated period of time, measure the volume of solution in each reservoir using the syringe. This allows one to calculate the volumetric flow rate from each reservoir.

4. If no flow is observed clean the microreactor as discussed above.

5. Analysis for product using a suitable technique such as GCMS or HPLCMS.

6. Rinse microreactor with small quantity of solvent and repeat reactions as necessary.

4. Product Purification in a Microreactor (29)

As mentioned above, peptide bond forming reactions within the micro reactor generally produce the dipeptide contaminated with only residual amounts of amine. Consequently, an initial study was to investigate the electrophoretic separation of peptide **3** from an equal concentration of amine **2**.

A solution of dipeptide **3** (25 μl, 0.1 M) and amine **2** (25 μl, 0.1 M) dissolved in anhydrous DMF were premixed and placed in reservoir G (ground electrode) of a microreactor (**Fig. 4.8**). Anhydrous DMF (25 μl) was placed in reservoirs A and an external voltage was applied in order to induce electrokinetic flow of the dipeptide **3** from the ground reservoir G to reservoir A (separated by 3 cm).

Fig. 4.8. Microreactor design for separation.

When various voltages (**Table 4.1**) were applied across the reservoirs, the peptide **3** was found to move from the ground reservoir G to reservoir A, with a corresponding decrease in the volume of DMF in reservoir A and increase in G. We therefore postulate that the DMF is moving from A to G under electro-osmotic flow whilst the peptide **3** moves in the opposite direction

Table 4.1
HPLC analysis of 1 and 2 at various voltages

Entry	Voltage B (V)	Area (%)	
		Amine	Peptide
1	800	–	100
2	900	–	100
3	1,000	–	100
4	1,100	–	100
5	1,200	–	100
6	1,300	–	100
7	1,400	–	100
8	1,500	–	100

via its electrophoretic mobility. The analysis illustrates that peptide **3** has moved to reservoir A (at the higher voltage), which leads to preconcentration of the amine **2** in reservoir G (ground), which could potentially be recycled in subsequent reactions.

Having demonstrated that it was possible to separate the desired peptide **1** from the amine **2**, a reactor was designed to enable the synthesis of the peptide **3** by reaction of the amine **2** with the pentafluorophenyl (PFP) ester **1**, followed by a subsequent separation within a single device (**Fig. 4.9**).

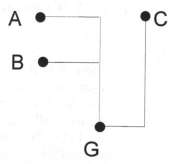

Fig. 4.9. Microreactor design for synthesis and separation.

A solution of amine **2** (50 μl, 0.1 M) in DMF was added to reservoir A and a solution of ester **3** (50 μl, 0.1 M) in DMF was placed in reservoir B and anhydrous DMF (20 μl) was placed in the ground reservoir G and reservoir C. A continuous voltage of 1,000 V and 900 V was applied for 10 min across reservoirs A and B, respectively, to collect the dipeptide product **1** in reservoir

G; HPLC analysis (using a Phenomenex Jupiter HPLC column) indicated that the peptide was produced in greater than 92% purity at the ground reservoir. The dipeptide was then electrophoretically mobilised from the ground reservoir G to reservoir C by applying 1,500 V at reservoir C for a further 10 min to collect the pure dipeptide (**Table 4.2**), which showed no trace of any amine **2** or pentafluorophenyl ester **1** (29).

Table 4.2
HPLC analysis of reservoirs G and C

		Reservoir C		
Entry	Peptide conversion/%	% 2	% 3	% 1
1	93	0	0	100
2	92	0	0	100
3	96	0	0	100
4	95	0	0	100

5. Notes

1. The EOF fluid velocity v_{eof} is given by Equation [**1**]

$$v_{\mathrm{eof}} = \frac{E\varepsilon\varepsilon_0\zeta}{\eta} \qquad [1]$$

   where E is the electric field (voltage divided by electrode separation), ε is the relative dielectric constant of the liquid, ε_0 is the permittivity of free space, ζ is the zeta potential of the channel wall–solution interface and η is the liquid viscosity. For EOF of organic solvents the two key parameters are the relative dielectric constant of the liquid ε and the viscosity η of the solvent. Consequently, the more polar the solvent and the lower the viscosity, the higher the flow rate will be.

2. Electroosmotic flow is generated using a power supply with multiple electrodes, produced by Kingfield Electronics, Sheffield, UK. LabVIEW software is used to deliver the correct voltage sequences to the electrodes in reservoirs relative to the product reservoir, set to ground voltage. The control system used at Hull also enables computer logging of all voltages and currents during a run.

Acknowledgements

The majority of the work described in this article was funded by Novartis, where Dr. Esteban Pombo-Villar is acknowledged for his support.

The authors wish to note that the field of micro reactors has developed significantly since this chapter was written. For more detailed information the authors suggest that the reader consults more recent reviews:

1. C. Wiles and P. Watts, *Eur. J. Org. Chem.*, 2008, 1655.

2. B. Ahmed-Omer, J. C. Brandt and T. Wirth, *org. Biomol. Chem.*, 2007, **5**, 733.

3. B. P. Mason, K. E. Price, J. L. Steinbacher, A. R. Bogdan and D. T. McQuade, *Chem. Rev.*, 2007, **107**, 2300.

References

1. Bradley, D. (1999) Chemical reduction. *European Chemist* **1**, 17.

2. Fletcher, P. D. I. and Haswell, S. J. (1999) Downsizing synthesis. *Chem. Br.* **35**, 38.

3. Cowen, S. (1999) Chip service. *Chem. & Ind.* 2 August, 584.

4. McCreedy, T. (1999) Reducing the risks of synthesis. *Chem. & Ind.* 2 August, 588.

5. Barrow, D., Cefai, J. and Taylor, S. (1999) Shrinking to fit. *Chem. & Ind.* 2 August, 591.

6. Cooper, J., Disley, D. and Cass, T. (2001) Lab-on-a-chip and microarrays. *Chem. & Ind.* 15 October, 653.

7. Haswell, S. J., Middleton, R. J., O'Sullivan, B., Skelton, V., Watts, P. and Styring, P. (2001) The application of micro reactors to synthetic chemistry. *Chem. Commun.* 391.

8. Matlosz, M., Ehrfeld, W. and Baselt, J. P. (eds.) (2002) *IMRET 5: Proceedings of the Fifth International Conference on Microreaction Technology.* Springer, Berlin.

9. Ehrfeld, W., Hessel, V. and Löwe, H. (2000) *Microreactors: New Technology for Modern Chemistry.* Wiley–VCH. Weinheim.

10. Overbeek, J. Th. G. (1952) in *Colloid Science*, ed. H.R. Kruyt, Elsevier, Amsterdam, Chap. V, 195.

11. Rice, C. L. and Whitehead, R. (1965) Electrokinetic flow in a narrow cylindrical capillary. *J. Phys. Chem.* **69**, 4017.

12. Hunter, R. J. (1981) *Zeta Potential in Colloid Science.* Academic Press, London.

13. Jednacak, J., Pravdic, V. and Haller, W. (1974) The electrokinetic potential of glasses in aqueous electrolyte solutions. *J. Colloid Interface Sci.* **49**, 16.

14. McCreedy, T. (2000) Fabrication techniques and materials commonly used for the production of micro reactors and micro total analytical systems. *TrAC* **19**, 396.

15. McCreedy, T. (2001) Rapid prototyping of glass and PDMS microstructures for micro total analytical systems and micro chemical reactors by microfabrication in the general laboratory. *Anal. Chim. Acta* **427**, 39.

16. Haswell, S. J. (1997) Development and operating characteristics of micro flow injection analysis systems based on electroosmotic flow. *Analyst* **122**, 1R.

17. Freemantle, M. (1999) Downsizing chemistry. *C&EN* 22 February, 27.

18. Harrison, D. J. and van den Berg, A. (eds.) (1998) *Proceedings of the Micro Total Analytical Systems '98 Workshop.* Kluwer Academic Press, Dordrecht.

19. van den Berg, A. and Lammerink, T. S. J. (1998) Micro total analysis systems: microfluidic aspects, integration concept and applications. *Topics Curr. Chem.* **194**, 21.

20. Woolley, A. T., Sensabaugh, G. F. and Mathies, R. A. (1997) High-speed DNA genotyping using microfabricated capillary array electrophoresis chips. *Anal. Chem.* **69**, 2181.

21. Jacobsen, S. C. and Ramsey, J. M. (1996) Integrated microdevice for DNA restriction fragment analysis. *Anal. Chem.* **68**, 720.

22. Schmalzing, D., Adourian, A., Koutny, L., Ziagura, L., Matsudaria, P. and Ehrlich, D. (1998) DNA sequencing on microfabricated electrophoretic devices. *Anal. Chem.* **70**, 2303.

23. Spence, D. M. and Crouch, S. R. (1997) Factors affecting zone variance in a capillary flow injection system. *Anal. Chem.* **69**, 165.

24. Service, R. F. (1998) Microchip arrays put DNA on the spot. *Science* **282**, 396.

25. Fletcher, P. D. I., Haswell, S. J., Pombo-Villar, E., Warrington, B. H., Watts, P., Wong, S. Y. F. and Zhang, X. (2002) Micro reactors: principles and applications in organic synthesis. *Tetrahedron* **58**, 4735.

26. Watts, P., Wiles, C., Haswell, S. J., Pombo-Villar, E. and Styring, P. (2001) The synthesis of peptides using micro reactors. *Chem. Commun.* 990.

27. Watts, P., Wiles, C., Haswell, S. J. and Pombo-Villar, E. (2002) Solution phase synthesis of β-peptides using micro reactors. *Tetrahedron* **58**, 990.

28. Watts, P., Wiles, C., Haswell, S. J. and Pombo-Villar, E. (2002) Investigation of racemisation in peptide synthesis within a micro reactor. *Lab. Chip* **2**, 141.

29. George, V., Watts, P., Haswell, S. J. and Pombo-Villar, E. (2003) On-chip separation of peptides prepared within a micro reactor. *Chem. Commun.* 2886.

Chapter 5

The Electroosmotic Flow (EOF)

Gary W. Slater, Frédéric Tessier, and Katerina Kopecka

Abstract

Controlling and manipulating liquids and analytes at the sub-millimeter scale is a challenge that frequently requires new methods to be developed. Indeed, scaling-down of traditional macroscopic ideas often fails. For instance, pumping liquids using pressure differences is often impractical and counterproductive because the resulting parabolic flow profile deforms sample zones. As the size of the system shrinks, the surface-to-volume ratio increases and interfacial effects become dominant. This actually opens new possibilities since the phenomenon of electroosmotic flow (EOF), wherein a fluid is made to move relative to a stationary charged boundary, can then be exploited to design efficient microfluidic devices. In this chapter, we review the fundamental principles of EOF as well as some of the methods used to coat channel walls and reduce the impact of EOF in situations where it would be unfavorable for the device performance.

Key words: Debye length, diffusion, electrical double-layer, EOF (electroosmotic flow), friction, screening.

1. Introduction

The electroosmotic flow (EOF) is a common phenomenon that can be both a nuisance and a tool. This chapter focuses on notions that can help researchers understand the fundamentals of EOF, analyze experimental data, and possibly design optimization strategies. It is undeniably theoretical in nature, but we believe that in order to exploit EOF and microfluidics, one must first be familiar with EOF. When a salt is present in a solution, an electric field makes the anions and cations move in opposite directions, thus generating an electric current but no liquid flow. In the case of EOF, however, one of the ionic species is fixed to a wall while the other moves in the solution in response to the applied field. The moving ions transfer their momentum to the liquid and, in the

M.P. Hughes, K.F. Hoettges (eds.), *Microengineering in Biotechnology,* Methods in Molecular Biology 583,
DOI 10.1007/978-1-60327-106-6_5, © Humana Press, a part of Springer Science+Business Media, LLC 2010

process, drag the rest of the liquid with them. Since most of the free ions reside near the charged surfaces due to electrostatic attraction, the EOF starts close to the walls, but it is quickly transferred to the whole solution via frictional forces. Unlike pressure-driven flows, EOF tends to provide plug flows in well-controlled geometries, a great advantage for separation systems. However, it is often preferable to suppress EOF because it may not be constant along the capillary, and the resulting flow gradients may affect resolution. **Section 2** below examines the basic theory of EOF, while **Section 3** describes some of the wall coating methods that are used to control EOF. Finally, **Section 4** suggests a list of readings that can complement this chapter.

2. Theory of EOF

The best known case where EOF plays a role is that of glass fluidic systems. The inner wall of a fused silica capillary or channel is negatively charged when in contact with standard aqueous buffers. The counterions are attracted to the wall and form a thin layer of cations of thickness $\lambda_D \cong 1$–50 nm, the Debye length. In the presence of an electric field E, the diffuse part of this layer moves and drags the liquid toward the cathode. This EOF can be described in terms of a mobility $\mu_{eof} = v_{eof}/E$, where v_{eof} is the steady-state bulk EOF velocity. It is possible to derive an expression for μ_{eof} from the electric permittivity ϵ, the solvent viscosity η, and the zeta potential ζ (the electric potential near the wall, at the plane of shear) (1):

$$\mu_{eof} \equiv \frac{v_{eof}}{E} = \frac{\epsilon \zeta}{\eta} \qquad [1]$$

There is an important hurdle, however, in predicting μ_{eof} for a given system. Indeed, the actual value of ζ (typically 1–100 mV) is not easily determined from first principles as it depends on the buffer's ionic strength and pH in non-trivial ways (2, 3). Therefore, one often uses the Helmholtz-Smoluchowski relation $v_{eof} = \epsilon \zeta E/\eta$ to measure ζ under suitable conditions. Since μ_{eof} specifies the *relative* motion of the electrolyte and the solid wall, it also serves as the electrophoretic mobility of a solid particle immersed in the electrolyte if λ_D is small compared to the size of the object (4).

The microscopic physics of EOF is generally described using the electrical double-layer (EDL) model, in which the counterions are pictured as forming two distinct layers near the solid wall (**Fig. 5.1**). The first one corresponds to the surface charged groups and the counterions transiently bound to them via electrostatic interactions; it is variably called the *Stern, Helmholtz,*

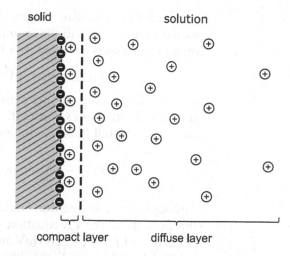

Fig. 5.1. Schematic representation of the accumulation of counterions in an electric double layer near a negatively charged solid wall. The wall charge typically arises from the chemical dissociation of surface groups, e.g., the deprotonation of silanols in the case of silica surfaces. Ions in the compact layer are essentially immobilized, while in the diffuse layer they are free to undergo diffusion. The net charge density in the diffuse layer is highest near the wall and decreases away from the surface to eventually reach the bulk value. The converse is true for the coions (not depicted here).

compact or *inner* layer, and it has a typical thickness of one ionic diameter to account for the finite size of the surface groups and bound ions (2). In the second layer, counterions are free to diffuse and the charge density and electric potential profiles are regulated by a Boltzmann distribution, i.e., by the competition between electrostatics and thermal motion. The characteristic thickness of this diffuse layer is a function of the bulk electrolyte concentration. The ζ potential is effectively the potential at the interface of the compact and diffuse layers, where the EOF shear takes place.

2.1. Charge Distribution

With this general EDL model in mind, we can work out practical expressions for the charge density profile $\rho(r)$ beyond the Stern layer. The first fundamental equation we need is the Poisson equation, which relates $\rho(r)$ to the electric potential $\phi(r)$ at position r:

$$\nabla^2 \phi(\mathrm{r}) = -\frac{\rho(\mathrm{r})}{\varepsilon} \qquad [2]$$

If k refers to one of N ionic species present in the solution, each of valence z_k (including sign), and if we denote by $n_k(r)$ the ionic number density at r, the local charge density is given by

$$\rho(r) = \sum_{k=1}^{N} e z_k n_k(r), \qquad [3]$$

where e is the elementary electric charge. The relation between the local potential $\phi(r)$ and the (equilibrium) number concentrations of diffusing ions is provided by a Boltzmann relation:

$$n_k(r) = n_{k0} \exp\{-ez_k\phi(r)/k_BT\}, \qquad [4]$$

where n_{k0} is the bulk number density of ion k, T is the temperature and k_B is Boltzmann's constant. Together, these equations lead directly to the full Poisson–Boltzmann (PB) equation:

$$\nabla^2\phi(r) = -\frac{e}{\varepsilon}\sum_{k=1}^{N} n_{k0}z_k \exp\{-ez_k\phi(r)/k_BT\}. \qquad [5]$$

Although one can question some of the implicit assumptions in this derivation, the PB equation is generally considered accurate (Equation [1]) for $\zeta < 200\,$mV and for electrolyte concentrations below 1 M. This non-linear differential equation has no general closed-form solution and even in the simplest cases of only one or two ionic species, it remains non-trivial to solve for $\phi(r)$ and $\rho(r)$. To retain a tractable equation, we must resort to the Debye–Hückel linearization scheme:

$$\nabla^2\phi(r) \approx \sum_{k=1}^{N} \frac{n_{k0}z_k^2 e^2}{\varepsilon k_B T}\phi(r) = \kappa^2\phi(r). \qquad [6]$$

(here, we also assumed that we have a neutral bulk electrolyte: $\Sigma n_{k0}z_k = 0$). This weak potential approximation is reasonable if $e\phi/k_BT < 1$ across the entire system (since $k_BT/e \approx 25$mV at room temperature, this condition is often violated). The parameter

$$\kappa \equiv \lambda_D^{-1} = \sqrt{\sum_{k=1}^{N} \frac{n_{k0}z_k^2 e^2}{\varepsilon k_B T}} \qquad [7]$$

measures the thickness of the diffuse layer. For water at room temperature, the Debye length is

$$\lambda_D \cong \frac{0.3}{\sqrt{I}}\,nm, \qquad [8]$$

where the ionic strength I is defined in terms of the concentration c_k of ion k (in mol/L):

$$I \equiv \frac{1}{2}\sum_{k=1}^{N} c_k z_k^2. \qquad [9]$$

We now examine a specific geometry.

2.2. A Cylindrical Capillary

Clearly, the solution of Equation [6] depends on the system geometry. Let us consider a cylindrical tube of internal radius a (i.e., the distance between its center and the boundary of the Stern layer) (5), a useful model for capillaries and microchannels. In fact, for a system of size a, the dimensionless product κa (called the *electrokinetic radius*) and the ζ potential completely specify the EOF. If r is the radial distance, the boundary conditions $\phi(a) = \zeta$ and $d\phi/dr|_{r=0} = 0$ lead to the solution

$$\phi(r) = \zeta \frac{I_0(\kappa r)}{I_0(\kappa a)} \qquad [10]$$

where I_0 is the modified Bessel function of the first kind. Equations [2] and [10] then give

$$\rho(r) = -\varepsilon \nabla^2 \phi(r) = -\varepsilon \kappa^2 \phi(r) = -\varepsilon \kappa^2 \zeta \frac{I_0(\kappa r)}{I_0(\kappa a)}. \qquad [11]$$

Note that this solution is self-consistent only if $\kappa a \gg 1$, i.e., if the Debye layer does not affect the charge distribution at the centre ($r = 0$). In other words, the dissociation of surface groups releases counterions in the solution, but the linearization process used in Equation [6] demands that this number be negligible compared to the number of ions already in the solution (e.g., from the added salt), thus maintaining "bulk" electroneutrality. This is probably invalid at the nanoscale because the large surface-to-volume ratio, but it is certainly sound at the microscale and above. When it is not valid, we must solve the PB equation with a self-consistent numerical scheme for both $\rho(r)$ and $\phi(r)$.

2.3. The Flow Profile

We can substitute expression 11 for $\rho(r)$ into the Navier–Stokes (NS) equation in order to obtain the radial profile of the resulting EOF axial fluid velocity $v(r)$ (5). If the fluid is Newtonian and incompressible, and if there is no pressure gradient along the channel, the steady-state NS equation for the velocity $v(r)$ in cylindrical coordinates assumes the form

$$\frac{\partial^2 v(r)}{\partial^2 r} + \frac{1}{r}\frac{\partial v(r)}{\partial r} = -\frac{E}{\eta}\rho(r). \qquad [12]$$

This can readily be integrated to yield the EOF velocity profile in the steady-state regime. If we consider the standard no-slip boundary condition $v(a) = 0$, we obtain

$$v(r) = -\frac{\varepsilon \zeta E}{\eta}\left\{1 - \frac{I_0(\kappa r)}{I_0(\kappa a)}\right\}. \qquad [13]$$

The ratio $I_0(\kappa r)/I_0(\kappa a)$ thus controls both the EOF velocity and charge density profiles, as shown in **Fig. 5.2** for several values of κa (the cases for which this ratio is substantial at $r = 0$ violate our initial assumptions, as mentioned above). When $\kappa a \gg 1$, the ratio is significant only in the Debye layer of thickness $\lambda_D = 1/\kappa$ near the wall; this corresponds to a fraction $2/\kappa a$ of the total tube cross-section. This is where the EOF is initially generated, but the steady-state flow is actually flat in the central part of the capillary, with a magnitude given by Equation [1].

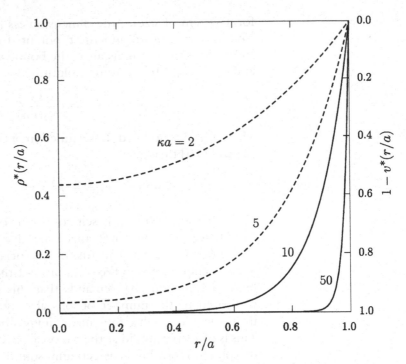

Fig. 5.2. The ratio $I_0(\kappa r)/I_0(\kappa a)$, for different values of the electrokinetic radius κa. According to Equations [12] and [14], this ratio gives the normalized charge density $\rho^* = -\rho/\varepsilon\kappa^2\zeta$ (*left axis*) and the normalized EOF steady-state velocity $\nu^* = -\eta\nu/\varepsilon\zeta E$ (*right axis*, reserved). The dashed curves indicates that the value of κa is too small to yield a self-consistent solution, as discussed in the text.

The plug-like flow is one of the most remarkable features of the EOF. We should point out that the expressions above are in their form very similar to those for a planar geometry. Indeed, if we replace the ratio $I_0(\kappa r)/I_0(\kappa a)$ by the exponential $exp(-\kappa x)$ in Equations [11] and [13], where x is now the distance to the closest charged plane, we obtain the appropriate functional forms for the charge density and fluid velocity near a flat wall. We can obtain the mean EOF velocity in the capillary by integrating the expression for $v(r)$ across the capillary section and dividing by the cross-sectional area:

$$\bar{v}_{eof} = -\frac{\varepsilon\zeta E}{\eta} \times \left\{ I - \frac{2}{\kappa a}\frac{I_1(\kappa a)}{I_0(\kappa a)} \right\} \cong -\frac{\varepsilon\zeta E}{\eta}$$

$$\times \left\{ 1 - \frac{2}{\kappa a} + \frac{1}{\kappa^2 a^2} + \cdots \right\} \kappa a \gg 1 \qquad [14]$$

The mean EOF velocity is thus given by Equation [1], with the expected leading correction factor $2/\kappa a$.

Analyte transport using the EOF is very efficient. If $\kappa a \gg 1$, the velocity is essentially uniform across the channel cross-section and the broadening of the analyte zone is controlled entirely by the analyte diffusion coefficient D. The lower value of the flow velocity in the vicinity of the wall leads to a small increase of the effective axial diffusion coefficient D_{eff} given by (6):

$$\frac{D_{eff}}{D} = 1 + \frac{1}{2}\left(\frac{Pe}{\kappa a}\right)^2 - \frac{11}{2}\left(\frac{Pe}{\kappa u}\right)^3 + \cdots \quad \kappa a \gg 1 \qquad [15]$$

where $Pe \equiv v_{eof}a/D$ is the Peclet number. The increased band broadening is a second order effect that is significant only for very high ratios $Pe/\kappa a$ not generally seen in microfluidic applications.

2.4. Transient Flow Time Scale

The previous derivation dealt with the steady-state EOF, but also intriguing is the transient regime that follows the application of an electric field to a stationary fluid. An acceleration term $\partial v/\partial t$ must then be added to the NS equation. Interestingly, the time-dependent solution for a slit geometry can be expressed (7) as a series of modes (n) decaying as e^{-t/τ_n}. The related time constants follow the simple law $\tau_n = \tau_1/n^2$, and the longest transient time ($n = 1$) is given by

$$\tau_1 = \frac{a^2 \delta}{\pi^2 \eta} \qquad [16]$$

where δ is the fluid density. The transient time thus increases like $\sim a^2$, but decreases when the solvent viscosity η increases. For water at room temperature, this gives $\tau_1 \approx (a/\mu m)^2 \times 0.1$ μs, a negligible time for most microfluidic applications. The role of the viscosity is interesting: while an increase of η reduces the steady-state EOF velocity, the latter is then reached sooner. This is due to the fact that the counterions need the shear friction (i.e., the viscosity) to transmit their momentum to the whole fluid. Note also that τ_1 is independent of other EOF parameters, such as the ionic strength or the valence of the electrolyte, the pH, the ζ potential, or even the field E.

2.5. Effect of pH and Ionic Concentration

The ζ potential determines the magnitude of the EOF, but its value depends on the chemistry of the solid–fluid interface. The details of this dependence are quite involved because the surface charge and the ζ potential are implicitly related to one another by relations involving parameters such as the ionic strength, the pH, the density of chargeable sites on the surface (which can be history-dependent) and their dissociation constant (the pK value), as well as the empirical Stern layer capacitance (2, 3). The pH, for instance, directly modulates the rate of dissociation of surface groups. For example, consider an aqueous buffer and a silica surface. Silanol (SiOH) groups on the surface behave as weak acids and remain protonated in acidic pH, but gradually dissociate to generate negative siloxy groups (SiO⁻) as the pH increases toward alkaline conditions (3, 8, 9). The pK value of SiOH groups is ~ 7, and in practice the fraction of negatively charged silanol groups becomes significant at a pH of about 2 and increases with pH to reach saturation around 9. The impact of the electrolyte concentration on ζ can be analyzed in simple terms if we neglect the presence of the Stern layer (i.e., ζ corresponds to the bare surface potential). The surface charge density σ is then simply given by Equation [11] and the overall electroneutrality condition:

$$\sigma = -\frac{1}{a}\int_0^a \rho(r)rdr = \frac{I_1(\kappa a)}{I_0(\kappa a)} = \varepsilon\kappa\zeta\left\{1 - \frac{1}{2\kappa a} - \frac{1}{8\kappa^2 a^2} - \cdots\right\}. \quad [17]$$

Increasing the electrolyte concentration causes κ to increase; therefore, the ζ potential must then decrease if the charge σ is fixed. Hence, a higher salt concentration generally implies a smaller ζ and a weaker EOF. It must be remembered, however, that this is a simplified picture and that in general both the surface charge and the surface potential vary with the solution conditions.

2.6. Conclusion

The discussion above is in part specific to particular scenarios, but nevertheless it does clarify the origin and the influence of the parameters typically involved in EOF problems. It is important to keep in mind the limits of applicability of the solutions, especially in the context of emerging sub-micron fluidic devices where one may have $\lambda^D \approx a$. However, for water at 300 K and a symmetric univalent electrolyte concentration of 0.1 M, $\lambda^D \approx 1$ nm, so the theory remains useful for a wide range of practical applications. In fact, other phenomena may dominate the physical picture at such small scales. For example, the discrete nature of the solvent and the consequent formation fluid layers over a range of many molecular radii (2) may affect the EOF picture in a significant way. In situations beyond the validity range of the continuous EDL model, we can resort to numerical techniques. This can take the form of a numerical integration of the PB and NS equations, which has the advantage of being amenable to practically any geometry, e.g., channel turns and cross junctions. Molecular dynamics computer simulations, in which the trajectory of every solvent molecule and ion is calculated explicitly, can also yield a very detailed, even atomistic picture of EOF at small scales (at an increased cost in computational time). Note that numerical schemes bear their own limitations in terms of underlying assumptions, which must be assessed carefully before a definite connection with actual experiments is established.

Using EOFs has several advantages. One of them is that even low surface charges lead to significant flows. In fact, it is interesting to note that for a given surface charge density, the magnitude of the EOF is independent of the channel radius a (the transient time, however, is a strong function of the radius). Consequently, the net flow rate produced by the EOF is simply proportional to the channel cross-section ($\sim a^2$), while the flow produced by a fixed pressure drop is proportional to $\sim a^4$, which means a must faster drop at small scales (10). Also, because all analytes are affected by EOF, a strong EOF often allows one to use a single detection point for all analytes, regardless of the sign and magnitude of their charge.

The main problem with EOF resides with its lack of reproducibility. For instance, the chemistry of the surface is affected by the history of the system. Furthermore, axial surface charge gradients give rise to pressure differences because the fluid is incompressible; this means that an unwanted Poiseuille flow is generated, which leads to strong axial dispersion. Non-uniform surface charges can also generate non-axial flows (such as rolls) with unpredictable effects on the separation (although such flows may help with sample mixing (10)). We should also add that Joule heating can lead to radial gradients and increased zone broadening. For these reasons, many applications require a suppression of the EOF. This is examined briefly in the next section.

3. Control and Modification of EOF

Electrophoresis in capillaries and microchannels offers an advantageous combination of speed and efficiency. However, the presence of bare walls often limits the number and types of analytes that can be separated. Generally, both the analyte–wall interactions and the EOF must be reduced or even eliminated using a robust coating process. Polymer coatings, either covalent or dynamically adsorbed, are thus widely used to modify the inner surface of the capillary wall or microchannels.

3.1. Covalent Coatings

The modification of the capillary wall by covalently bound polymers for the separation of proteins was first demonstrated by Hjertén in 1985 (11). Such treated capillaries can offer a wide range of accessible pH and can withstand any of the aggressive rinsing solutions that might be applied for regeneration of the capillary surfaces. An initial derivatization of the inner surface of the capillary/microchannel wall is usually required for creating a covalently linked polymer coating. The procedures for the silanization of walls, which include the preconditioning of the silica surface, the type of silanization solvent, and the silane agent itself vary widely. For example, coatings prepared using toluene as the silanization solvent were shown to have superior stability compared to those prepared either with 95% ethanol or with an aqueous solvent containing an acetic acid catalyst. Although mono-chlorsilanes can provide denser surface coverage, di- and tri-chlorsilanes improve the stability of the silane coating. Unfortunately, these procedures are laborious and time consuming.

Furthermore, they are typically carried out as difficult-to-control in situ polymerization, thus affecting the reproducibility of the prepared coating (12).

3.2. Dynamic (Adsorptive) Coatings

Although numerous dynamic coating procedures exist, in most cases EOF is simply suppressed by flushing the capillary or microchannel with a low-viscosity dilute polymer solution, which obviates the need for organic solvents, high temperatures, and viscous solutions, all of which are required to produce covalently linked coatings. This is obviously a highly desirable situation, although one may sometimes question the stability of such coatings.

3.3. How do Coatings Work?

The suppression of EOF by covalently bound polymers has been attributed to at least three different effects: (i) the modification of the ζ potential via the elimination of ionizable SiOH groups on the fused-silica surface; (ii) the "shielding" of the remaining surface charged groups by the polymer (or some other additive); and (iii) the increase of the viscosity in the EDL region (13). At this point in time, we not have a good theoretical understanding of this phenomenon.

In the case of dynamic coatings, the surface, the polymer (plus its conformations), and the solvent, as well as the interactions between these components, directly affect polymer adsorption. The polymer-wall interactions may involve EDL effects, van der Waals forces, steric repulsion, hydration/solvation interactions (attractive/repulsive), polymeric bridging, and hydrophobic interactions (both attractive). However, a full understanding of the adsorption mechanism is still missing. In many cases, the adsorption behavior of the polymer is simply attributed to the hydrophobicity of the polymer chain. Although the origin of the hydrophobic force still remains controversial, its effect in molecular association problems has long been recognized (14).

The best fundamental study of EOF control via dynamic coating was published by Barron et al. (15). These authors synthesized a set of model polymers and copolymers based on N,N-dimethylacrylamide (DMA) and N,N-diethylacrylamide (DEA) in order to examine the critical factors that control the efficiency of dynamic coating. They showed that the contour length of the polymer affects its ability at suppressing the EOF. In the case of PDMA and PDEA, for instance, polymer chains consisting of 5,000 monomers reduce the EOF mobility by one order of magnitude compared to uncoated capillaries. Above 15,000 monomers, however, the EOF is reduced even more, to about 10^{-9} m^2(Vs)$^{-1}$, which is adequate for most applications. An interpretation is that adsorbed polymers form loops whose size must exceed that of the Debye layer in order to fully absorb the momentum of the counterions. This implies that a minimum contour length is needed. Further studies will be required to establish the precise mechanism responsible for this effect. The residual EOF mobility can be measured in capillaries by the method of Williams and Vigh (16) or in microchannels by the method of Huang et al. (17).

3.4. Some of the Main Coating Polymers

Table 5.1 presents some polymers used for EOF control via dynamic polymer wall coating. For bioseparations, PDMA and PVP are perhaps the most widely used. The main disadvantage of

Table 5.1
Overview of polymers used for EOF control: dynamic coating (23, 24)

1a. Cationic polymers – (typically used for positively charged proteins, peptides, drugs, etc.)	
Polyamine	
Poly(dimethyldiallylammonium chloride), PDADMA	Improves the resolution at neutral pH.
Poly(ethyleneimine), PEI	Excellent stability for pH between 3 and 11.
Polybrene	Less reproducible than PDADMA.
Poly(arginine), PA	vs. Polybrene: higher efficiencies, but less stable.
Oligoamins	Used with acidic pH.
Cationic cyclodextran	
1b. Anionic polymers – (used for the efficient separation of acidic proteins and amino acids)	
Dextran sulfate, DS	pH-independent EOF between pH 2 and 11.
Poly(vinyl sulfonic acid)	Less efficient than DS, as the hydrophilicity of the coating plays an important role for protein separation.
2. neutral polymers – (widely used for suppression of EOF for DNA and other biomolecule analysis)	
Polysaccharides	
Cellulose acetate, cellulose triacetate	Efficient separation of basic proteins.
Cross-linked hydroxypropylcellulose	
Hydroxypropylmethylcellulose, HPMC	Separations of H1 subtypes at low pH.
Hydroxyethylcellulose, HEC	Sieving agent for DNA restriction fragments.
Guaran	Buffer modifier to improve the separation of basic proteins and drugs. Does not form a stable coating.
Poly(vinyl alcohol), PVA	Separation of DNA restriction fragments. Effectively suppresses EOF below pH 8. Used with HEC or acrylamide matrixes, though PVA forms inhomogeneous networks after a few runs; binds more strongly to silica surface than HEC does.

(continued)

Table 5.1 (continued)

Poly(alkylene glycol) polymers	
Poly(ethylene oxide), PEO	For DNA analysis. More efficient for protein analysis than HEC and HPC. Capillaries must be washed before coating and between runs with hydrochloric acid. Unstable coating at higher pH.
Poly(vinylpyrrolidone)	Genotyping and sequencing applications.
Poly(dimethylacrylamide), PDMA	DNA sequencing. Excellent adsorptive properties.
Poly-N-hydroxyethylacrylamide, PHEA	Protein separations. pH ranging between 3 and 6.

most available polymers is their limited effectiveness for protein separations, especially under basic pH conditions. New acrylic polymers bearing oxirane groups were synthesized for the dynamic coating of capillaries, and they suppress EOF to a negligible value

Table 5.2
Overview of polymers used for EOF control: covalent coatings (25, 26)

Polyacrylamide, PA	Protein separations. Amino groups hydrolyze at high pH.
Poly(acryloylaminoethoxyethanol)	Life time at pH 8.5 is more than twice than that of PA.
PVA	EOF reduction (pH 3–10). Protein separations.
PEO	Tunable EOF, vary with the size of the PEG groups attached to the wall.
Poly(N-(acroylaminoethoxy)ethyl-β-D-glucopyranose	
Halo and alkoxy silanes, *e.g.* (a) 3-glycidyloxypropyltrimethoxysilane (b) 3-glycidyloxypropyltrimethoxysilane	(a) Separation of geometric isomers, inorganic and organic anions, polycarboxylates and polyphosphates; (b) Separation of ribonucleotides, peptides and proteins. EOF is greatly suppressed for pH 3–10.
Cross-linked polyamines	Behaves as anion exchanger.
Immunoaffinity chromatography	Protein analysis. Biospecific interactions, complementary behavior.
Molecularly imprinted polymers	Enantioseparations. Possesses binding characteristics that are comparable to the biological antibodies. Applicable to miniaturized systems.

(18). The adsorbed coatings were stable for hundreds of hours at high pH and temperature, and in the presence of 8 M urea, thus allowing efficient separation of acidic and basic proteins.

Recently, we have seen the synthesis of new types of polymers that can both coat walls and act as tunable separation matrices that change their properties in response to external stimuli (19). Liang et al. (20), for example, prepared a polymer which shows good self-coating abilities together with a temperature-dependent viscosity by grafting PEO to N-isopropylacrylamide. Such a polymer is indeed close to being the "ideal" matrix, i.e., a matrix that (i) can be easily pumped in and out of the channel, (ii) has good separation properties, and (iii) coat the walls (21).

In the case of covalent coatings, polyacrylamide and various copolymers of acrylamide are widely used because they exhibit excellent permanent coatings properties (**Table 5.2**).

We should add that the use of charged polymer layers for the suppression of EOF and analyte adsorption can also be considered. They are simple to produce and show excellent properties for the analysis of single-protein samples; however, they are not practical for the separation of highly complex mixtures. On the other hand, the use of polyelectrolyte multilayers is a promising approach to generate uniform surfaces with stable EOF properties on polymeric walls (chips).

Acknowledgments

K.K. would like to acknowledge the financial support of the Natural Sciences and Engineering Research Council of Canada (NSERC) and of NATO for a postdoctoral fellowship. GWS would like to thank NSERC for a Discovery Grant. F.T. would like to thank the University of Ottawa for an admission scholarship.

References

1. Russel, W. B., Saville, D. A. and Schowalter, W. R. (1989) *Colloidal Dispersions.* Cambridge University Press, New York.
2. Israelachvili, J. N. (1992) *Intermolecular and Surface Forces.* Academic Press, New York.
3. Behrens, S. H. and Grier, D. G. (2001) The charge of glass and silica surfaces. *J. Chem. Phys.* **115**, 6716–6721.
4. Viovy, J.-L. (2000) Electrophoresis of DNA and other polyelectrolytes: physical mechanisms. *Rev. Mod. Phys.* **72**, 813–872.
5. Rice, C. L. and Whitehead, R. (1965) Electrokinetic flow in a narrow cylindrical capillary. *J. Phys. Chem.* **69**, 4017–4023.
6. Datta, R. and Kotamarthi, V. R. (1990) Electrokinetic dispersion in capillary electrophoresis. *AICHE J.* **36**, 916–926.
7. Patankar, N. P. and Hu, H. H. (1998) Numerical Simulation of electroosmotic flow. *Anal. Chem.* **70**, 1870–1881.
8. Berhens, S. H. and Borkovec, M. (1999) Electrostatic interaction of colloidal surfaces with variable charge. *J. Phys. Chem. B* **103**, 2918–2928.

9. Burns, N. L., Emoto, K., Holmberg, K., Van Alstine, J. M., and Morris, J. M. (1998) Surface characterization of biomedical materials by measurement of electroosmosis. *Biomaterials* 19, 423–440.

10. Stone, H. A., Stroock, A. D., and Ajdari, A. (2004) Engineering flows in small devices: microfluidics towards a lab-on-a-chip. *Annu. Rev. Fluid Mech.* 36, 381–411.

11. Hjertén, S. (1985) High-performance electrophoresis – elimination of electroendoosmosis and solute adsorption. *J. Chromatogr.* 347, 191–198.

12. Albarghouthi, M. N., Stein, T. M., and Barron, A. E. (2003) Poly-N-hydroxyethylacrylamide as a novel, adsorbed coating for protein separation by capillary electrophoresis. *Electrophoresis* 24, 1166–1175.

13. Doherty, E. A. S., Meagher, R. J., Albarghouthi, M. N., and Barron, A. E. (2003) Microchannel wall coatings for protein separations by capillary and chip electrophoresis. *Electrophoresis* 24, 34–54.

14. Birdi, K. S. (1997) *Handbook of Surface and Colloid Chemistry.* CRC Press LLC, New York, pp. 559–600.

15. Doherty, E. A. S., Berglund, K. D., Buchholz, B. A., Kourkine, I. V., Przybycien, T. M., Tilton, R. D., and Barron, A. E. (2002) Critical factors for high-performance physically adsorbed (dynamic) polymeric wall coatings for capillary electrophoresis of DNA. *Electrophoresis* 23, 2766–2776.

16. Williams, B. A. and Vigh, G. (1996) Fast, accurate mobility determination method for capillary electrophoresis. *Anal. Chem.* 68, 1174–1180.

17. Huang, X. Gordon, M. J., and Zare, R. N. (1988) Current-monitoring method for measuring the electroosmotic flow rate in capillary zone electrophoresis. *Anal. Chem.* 60, 1837–1838.

18. Chiari, M., Cretich, M., Damin, F., Ceriotti, L., and Consonni, R. (2000) New adsorbed coatings for capillary electrophoresis. *Electrophoresis* 20, 909–916.

19. Sassi, A., Barron, A. E., Alonso-Amigo, M. G., Hion, D. Y., Yu, J. S., Soane, D. S., Hooper, H. H. (1996) Electrophoresis of DNA in novel thermoreversible matrices. *Electrophoresis* 17, 1460–1469.

20. Liang, D., Song, L., Zhou, S., Zaitsev, V. S., and Chu, B. (1999) Poly(N-isopropylacrylamide)-g-poly(ethyleneoxide) for high resolution and high speed separation of DNA by capillary electrophoresis. *Electrophoresis*, 20, 2856–2863.

21. Heller, Ch. (2001) Principles of DNA separation with capillary electrophoresis. *Electrophoresis* 22, 629–643.

22. Ghosal, S. (2004) Fluid mechanics of electroosmotic flow and its effect on band broadening in capillary electrophoresis. *Electrophoresis* 25, 214–228.

23. Horvath, J. and Dolnik, V. (2001) Polymer wall coatings for capillary electrophoresis. *Electrophoresis* 22, 644–655.

24. Righetti, P.G., Gelfi, C., Verzola, B., and Castelletti, L. (2001) The state of dynamic coatings. *Electrophoresis* 22, 603–611.

25. Liu, C. Y. (2001) Stationary phases for capillary electrophoresis and capillary electrochromatography. *Electrophoresis* 22, 612–628.

26. Liu, C. Y. (2001) Stationary phases for capillary electrophoresis and capillary electrochromatography. *Electrophoresis* 22, 612–628.

Further Reading

The present chapter is a simple introduction to the physics of EOF and the control of EOF via polymer coating. The following references are highly recommended: (4) is perhaps the best review of the electrophoresis of macromolecules; (10) is a superb review of electrokinetic phenomena in microfluidic devices and covers more materials that we could explore here; (13) is an extensive review of polymer coating; (22) is a great survey of some of the mathematical aspects of EOF.

Chapter 6

Microengineered Neural Probes for In Vivo Recording

Karla D. Bustamante Valles

Abstract

One of the great challenges facing medicine is the repair of the damaged nervous system. Due to the limited capacity of the central (and to a lesser extent the peripheral) nervous systems to regenerate, damage such as spinal cord injury can often result in permanent paralysis. Researchers are attempting to overcome nerve injury by devising methods of sensing neural activity either in the brain or in the spinal cord or peripheral nervous system. This information can act as a control mechanism for either muscle stimulators (e.g. for restoring limb function) or providing function in some other way (such as controlling a cursor on a computer screen). Ideally, sensing devices are implanted into the body, directly accessing the nervous system. Whilst great advancements have been made in implantable neural stimulators, sensing of neural activity has proven to be a more difficult task. This chapter describes how microengineered probes allow construction of neuron-sized neural interfaces for enhanced recording in vivo.

Key words: Nerve, stimulation, recording, neuron, microengineering.

1. Introduction

Neural probes are devices used to sense signals from the nervous system. Since the mid-1800s, scientists have tried to understand the operation of the nervous system. Many attempts to record neural activity have been made and different types of electrodes have been developed. However, it was not until the experiments of Hodgkin and Huxley from 1938 to 1952 that a satisfactory description of nerve signals was found, and they were recorded. As described by Hausser (1), four key developments were crucial for the development of neural recordings. The first, demonstrated by Cole and Curtis (2), was the discovery that action potentials are related to membrane conductance. After that, Hodgkin and Huxley (3) achieved the first intracellular recording of action potentials. Later

M.P. Hughes, K.F. Hoettges (eds.), *Microengineering in Biotechnology*, Methods in Molecular Biology 583,
DOI 10.1007/978-1-60327-106-6_6, © Humana Press, a part of Springer Science+Business Media, LLC 2010

on, Hodgkin worked together with Katz (4) to demonstrate that the sodium permeability causes the overshooting action potential. Finally, Hodgkin and Huxley (5) developed a voltage-clamp circuit to measure the ionic currents from the axon of a squid. These experiments, using single electrodes, permitted the study of the behaviour of single neurons. However, to understand the nervous system, and the complex processing that relies on vast neural connections, it is necessary to obtain information from many neurons at the same time. The purpose of modern neural probes is to record from as many individual neurons as possible without causing damage to the cells. Some of the essential characteristics for these modern devices are as follows:

- Size: Since it is required to record from single neurons, the size of the electrode has to be similar to that of the neurons, of the order of micrometres.

- Material: As the body's immune system will reject any foreign object, the electrode has to be of a material that does not damage or cause rejection and that does not become affected by the internal conditions within the body, i.e. it must be *biocompatible*.

- Insertion and attachment: The way the probe is inserted and attached to the nerve has to be of a manner that the nerve remains as intact as possible and the probe does not change position with movement.

To obtain a neural probe with these characteristics, many approaches have been investigated. Cuff electrodes, regeneration probes and penetrating microelectrodes are the main approaches to date. This chapter describes the development of implantable neural probes for recording from neurons in vivo.

2. Cuff Electrodes

A cuff electrode is a device that encircles the whole nerve (**Fig. 6.1**). It consists of an insulating sleeve placed around a length of nerve, with electrodes around it. The advantages of this technology are that the nerve remains intact during implantation and surgical implantation is not difficult (44). Although the implantation is not difficult, after surgery the nerve will swell and connective tissue will form where surgery was performed. These factors must be taken into consideration to avoid any neural damage. Possible displacement can be avoided by mechanical fixation. The main disadvantage of this technique is that the recordings are made of the sum of superficial neurons, as the electrode encloses the whole nerve. Signals from neurons inside

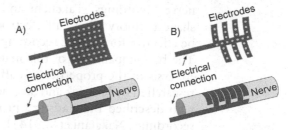

Fig. 6.1. Cuff electrodes: (a) half cuff electrode (43). (b) flexible interdigitating cuff electrode.

the centre of the nerve have very low amplitude at the surface, and may be lost in noise. Signal processing can be used to separate information from different neurons and to obtain information about neural coding; however, this is difficult to achieve. Other important restrictions are the material and the geometry, which have to be careful chosen since bending torques, compression to the nerves or disruption to nerves blood supplies can provoke severe nerve damage.

As described by Heiduschka and Thanos (44), the first electrodes were made of rigid materials and with very few electrodes. The electrodes were made of platinum foil (6) or platinum wire electrodes (7). These first models presented major problems, such as damaging the nerve due to mechanical displacement or causing ischemia due to pressure on the nerves. The foil electrode also had a problem with weak joints, since a weld is required between the joint and the lead-out cable. Platinum wire electrodes were usually sewn into the cuff wall. However, these electrodes were not likely to stay inside the wall of the cuff, and connective tissue would grow around them making then impossible to remove without damage to the nerves. To overcome these problems, new geometries and materials were developed. In the early 1980s, the use of platinum wire for electrodes and leads was often substituted by Teflon coated, multi-stranded stainless steel wires embedded into the cuff wall, which have been widely used for the remaining 20 years (e.g. (8)). The electrode is made from the same lead-out wire and as it is embedded into the cuff wall provides a mechanically flexibility. The introduction of a self-spiralling cuff design by Naples et al. (9) helped reducing the need for rigidity in the cuff wall and minimising the wall thickness. Klepinski (10) patented a chronically implantable cuff electrode with semi-rigid fingers extending orthogonally from the central spine, which are bent around the nerve. Sahin and Durand (11) reported an improved nerve cuff electrode using electric currents to slow the velocity of the action potentials, which creates an analogous effect of increasing the cuff length without longer intercontact delays. Their data suggested that this method could be used to improve the SNR of nerve recordings. Takeuchi and Shimoyama (12) developed a shape memory alloy cuff electrode, which has demonstrated to be effective for chronic recordings in a wireless system on insects. The biocompatibility of this material is still under investigation. Analysis of the properties of cuff electrodes has been extensively researched by Jezernik and Sinkjaer (13) from Aalborg University, who described the statistical properties of the whole nerve cuff recordings. Nakatani et al. (14) have developed methods to detect action potentials using cuff electrodes with a low signal to noise ratio, trying to address one of the main disadvantages of this type of electrodes compared to intrafascicular electrodes. These cuff electrodes were made of 75-μm-diameter stainless-steel wire and they were designed for bipolar recording. The processing of the data was not done in real-time, which would be necessary for real clinical applications. Jensen and co-workers are currently investigating the use of a triple electrode cuff electrode to provide motion feedback for functional electrical stimulations system by detecting joint rotation using recordings on sciatic nerves of rabbits (15).

Many types of cuff electrodes are already commercially available, but are mainly used for stimulation purposes. For example, the Vagus Nerve Stimulation System to help treat patients with epilepsy (Cyberonics Corp.); Bladder Control System from Vocate; NeuroControl Corporation control bladder and bowel voiding, or the Paralysed Hand Control System from FreeHand NeuroControl Corporation to help activate grasp in high-level quadriplegia. A more detailed review on cuff electrodes can be found in Hoffer and Kallesoe (16).

Although there have been great advantages in the design of cuff electrodes, they have many drawbacks when used for selective neural recordings. Whilst cuff-based stimulator electrodes are widely used, there are still no commercial devices which have been used for recording neural activity. One of the main drawbacks of the cuff type of sensor is selectivity.

3. Regenerating Electrodes

Regenerating electrodes are situated within the nerve. The probes are sieve-shaped and contain holes, with electrodes on the side of the holes or on the walls of these holes. The nerve is cut and the probe is inserted in the gap (**Fig. 6.2**). The neurons are then allowed to regenerate through the holes. Direct recordings can be made from the neurons and the probe is mechanically stable as the neurons grow through the holes and keep it in position. The main drawback of this method is the need to cut the nerve, thereby potentially causing irreparable damage, and the length of time it

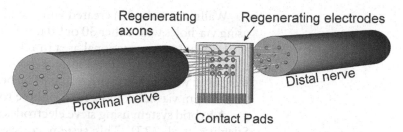

Fig. 6.2. Regenerating electrodes.

takes the nerve to regenerate. For example, Edell et al. (17) reported nerve recovery after 332 days in New Zealand rabbits. Akin et al. (18) achieved regeneration after 118 days on 21 out of 28 implants on rat nerves. These types of electrodes are used only on peripheral nerves because the central nervous system does not regenerate.

As described by Kovacs (19), the first work on constructing regenerating electrodes was undertaken by Frishkoff (unpublished) around 1965 with no successful results. Other authors continued researching this technique and it was not until 1974 that Mannard et al. (20) reported the first recording using nerve regeneration electrode units. The probes were made on an epoxy wafer and the holes were mechanically drilled. The first recordings using silicon-based regeneration electrodes were not made until the beginning of the 1980s. Edell et al. (17) reported the development of regenerating nerve electrodes using silicon bars, 40 μm wide and 200 μm apart in a comb configuration. These probes had 20×40 μm electrodes and were implanted in rabbit peripheral nerves. The results showed the ability of this design to record action potentials up to 600 μV in amplitude. Kovacs et al. (21) developed a new process for the fabrication of regenerating microelectrode arrays for peripheral and cranial nerve applications using BiCMOS processes (CMOS technology not involving high temperature process steps nor heavily doped etch stop layers). A typical probe had 100 μm^2 microelectrodes and 90×90 μm via-holes. These devices were implanted in the rat peroneal nerve and the frog auditory nerve; action potential signals were recorded from several microelectrodes in parallel. Another design of regeneration probes was developed by Akin et al. (18). This probe consisted of a silicon rim support, with a thickness of 15 μm, and different sizes of via-holes. The electrodes had areas from 100 to 2,000 μm^2 and were made of iridium. A silicon ribbon cable was integrated to connect the probe to the outside world. These electrodes were tested on peripheral taste neurons of rats. Regeneration was successful and electrophysiological recordings were obtained.

Wallman et al. (22) created and tested regenerated electrodes using via-holes with either 30 or 90 μm and varying the ratio of the total via-hole cross-sectional area in relation to the actual surface area of the electrode. Previously, recordings with 50 or 10 μm were used with unsuccessful results. The best results were achieved with 30 μm via-holes and a ratio of hole to surface area of 30%.

A hybrid system using sieve electrodes is being investigated by Stieglitz et al. (23). This system consists of a microfabricated flexible sieve polyamide device with microelectrodes and integrated biological cells. The device has 19 electrodes for stimulation and recording purposes and the adaptation of the implanted biological cells to the nerve stump has been demonstrated. Their future work will focus on the integration of cell containments with their microsystem.

This type of neural probes provides a close contact with the neurons, allowing a selective recording and stimulation besides mechanical stability. However, there is no guarantee of healing and regeneration through the probe. The lesion to the nerve caused in order to implant the probe will only heal under favourable conditions, which cannot always be controlled.

Cuff electrodes and regenerating-type electrode have been developed for many years now. There have been great advances; however, there are many drawbacks associated. Penetrating probes offer good compromise between cuff and regenerating-type electrodes. As described next, they allow a better selectivity than cuff electrodes as they are in direct contact with neurons but they are not as invasive as regenerative-type electrodes.

4. Penetrating Electrodes

The third type of device is a "needle like" probe carrying one or more electrodes. These probes penetrate inside the nerve and can have direct contact with one or more neurons. The first types of penetrating electrodes were simple insulated metal wires or glass filled pipettes with a metal contact inside (**Fig. 6.3**).

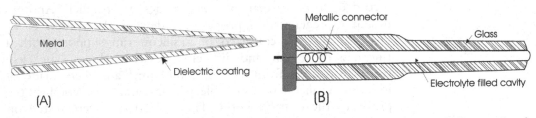

Fig. 6.3. First microelectrodes, modified from Ferris (45) (**a**) Cross section of a metal wire electrode. (**b**) Cross section of a glass electrode.

These wire or glass microelectrodes were used for the early neurophysiological experiments, for example, in the work of Hodgkin and Huxley mentioned before, and they are still in use today (24). Very fine metal tips could be created, sometimes up to 0.2 μm in diameter as reported by Grundfest et al. (25). A review of the first type of electrodes can be found in work by Kruger (26). These probes usually allow only single unit recordings; whilst this was very useful for the understanding of the nerve system, simultaneous recordings from many neurons are needed as the nervous system is made of large and complicated neural connections.

The use of photographic techniques was developed to help the creation of multielectrode microprobes. Gross et al. (27) developed a 36-electrode array with photoetching techniques for in vitro recordings. The contacts had 12 μm diameter and neural activity from several neurons was recorded. Pine (28) introduced two rows of electrodes 250 μm apart with electrodes surfaces of 8×10 μm produced by photoetching techniques. A review on microetching and metal deposition techniques can be found in Pickard et al. (29).

Another technique emerged in the early 1970s with the use of the semiconductor industry for the fabrication of microelectrodes. One of the first electrodes reported using silicon techniques was that by Wise et al. (30). They produced an array of gold electrodes supported on a silicon carrier. Interelectrode spacing was accurately controlled and tip diameters as small as 2 μm were produced. Successful recordings were obtained from cat cortex. Later, Wise and Angell (31) reported a multielectrode structure containing integrated junction-FET input stages. Prohaska et al. (32) obtained an eightfold microelectrode using thin-film technology with electrodes contact areas of 50×50 μm and separation of 300 μm. Since then, many microprobes with different characteristics (such as geometrical areas, materials and number of electrodes) have been fabricated. Gold, platinum and iridium have been widely used for recording neural activity. They all present good electrical and biocompatibility characteristics; however, if the probe is to be used for stimulation purposes, iridium has a higher current density capability. The area of the electrodes reported varied from 10 up to 2,500 μm^2. The different types of design include planar electrodes in a two-dimensional plane or electrodes with a three-dimensional form. A simple wire penetrating electrode would usually have three-dimensional recording capabilities, meaning it would be able to record from all its surroundings. Usually the silicon probes, will have electrodes in a two-dimensional plane, being less able to record action potentials that occur behind the plane of the electrode. To overcome this problem and still use the advantage of silicon probes, three-dimensional arrays have been fabricated using an array of planar shanks, or using electrodes needles in a brush-like array (e.g. **Fig. 6.4**).

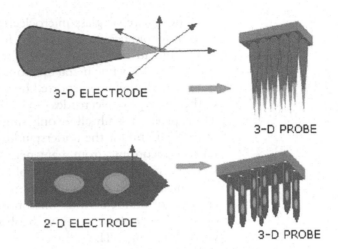

Fig. 6.4. Two- and three-dimensional neural probes.

The University of Surrey, in collaboration with the University of Southampton, has been working for more than 10 years in the development of 2D silicon microprobes (33). The probes were fabricated using a *bonded and etched-back silicon-on-insulator* wafer (BESOI), which permits the reduction of boron doping for the future implementation of on-site electronics. The probes used different sizes of electrodes allowing the investigation of the effects of diverse recording areas on the neural recordings. The probes also had different sizes of shank width, from 122 to 250 μm. These probes have successfully recorded neural activity from locust nervous system and the different designs permit the study of the different recording characteristics.

One of the groups that have worked for many years developing penetrating silicon probes is at the Centre for Neural Communication Technology under the direction of Anderson and Wise. Their work has been directed to the development of multichannel silicon substrate probes with one to four shanks and up to 16 (177 μm^2) iridium electrodes per shank. A diagram of one of these probes is found in **Fig. 6.5**. Some of the latest publications from this group report single-unit recordings from three-dimensional electrode arrays (**Fig. 6.6**) containing on-chip signal circuitry (34).

Another recent development is found in Xu et al. (35). They reported a metal microelectrode array (**Fig. 6.7**), similar to tungsten-wire probes or tetrode probes, but made using metal shanks and electrical insulation by conformal coating of Parylene-C film. The recording sites have precise opening sizes defined by

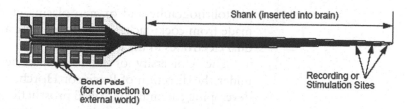

Fig. 6.5. CNCT, Schematic of a single shank acute probe. Picture courtesy Dr. D. Anderson.

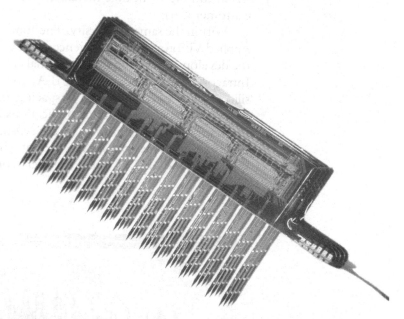

Fig. 6.6. 3-D probe array, courtesy Dr K. Wise. Reprinted with permission from Proceedings of the International Conference on Solid State Sensors and Actuators (Transducers 01), Munich. © 2001 IEEE.

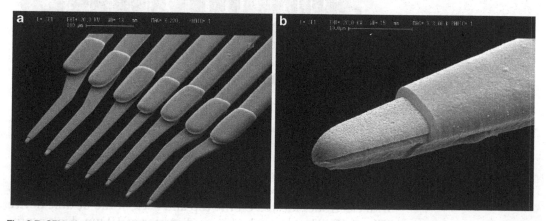

Fig. 6.7. SEM of metal microelectrode (a) View of the shank design; (b) close view of the recording tip. Pictures courtesy Dr. C. Xu.

photolithography and oxygen plasma etching. Recordings were made from cockroaches and signals were acquired from all individual electrodes in the probe.

The University of Utah, Department of Bioengineering, under the direction of Professor Horch, have worked extensively developing a neural-controlled prosthetic arm. This arm would be controlled by neural signals from motor nerves in the amputee' stump. To sense these nerve signals, flexible microwires were developed. Ti/W, Au and Pt were used to produce mechanically stable and highly flexible metallised fibres insulated with silicone elastomer (36).

Within the same University, The John Moran Laboratories for Applied Vision and Neural Sciences have worked for many years on the development of three-dimensional electrode arrays. The Utah Intracortical Electrode Array (UIEA) consists of 100 penetrating silicon microelectrodes with a spacing of 400 μm (**Fig. 6.8**). However, chronic recording applications are still limited. These probes have been implanted in cat sensory cortex for up to 13 months. In some cases evidence of chronic injury response was found and only 11 electrodes were connected to the exterior (37–39).

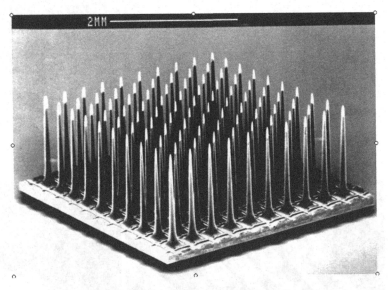

Fig. 6.8. The Utah Intracortical Electrode Array, courtesy Dr. R. Normann.

Stieglitz and Gross (40) recently reported a process technology for polyamide-based devices. These are flexible microprobes with electrodes located on the top and back of the flexible device (**Fig. 6.9**). The electrodes have a diameter of 10 μm. First prototypes were built and electrically tested; however, they have not yet

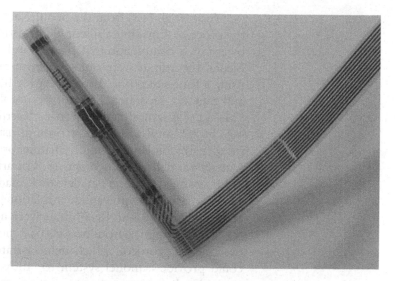

Fig. 6.9. Photograph of the flexible microprobe, courtesy of Dr. Koch.

been tested in any animal model. This work is done at the Fraunhofer-Institute Biomedical Engineering, Neural Prosthetics Unit, Ingbert Germany.

At the Toyohashi University of Technology, in Japan, Kawano et al. (41) have developed a micro-Si wire probe array with on chip circuits using a selective Au-Si$_2$H$_6$ vapour–liquid–solid growth method. Si probes with 160 μm in length and 3.5 μm in diameter at the tip were grown (**Fig. 6.10**). Simulated stimulation of neural activity was recorded to evaluate these prototypes, which could mean that these probes could be useful for real neural recordings.

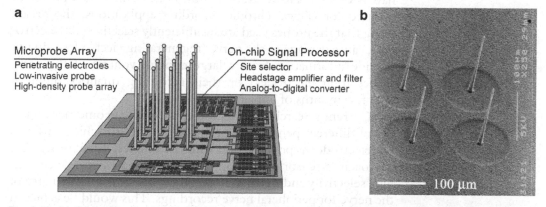

Fig. 6.10. (**a**) Schematic diagram illustrating an image of "smart neural probe chip," comprising a multichannel penetrating Si microprobe electrode array and with on-chip integrated circuit (IC) signal processors. The Si probe array was fabricated by selective growth of Si probes, which realised low neural invasiveness probes and high-density probe array with distribution similar to neurons. The greatest merit of this technology is that the microprobe array can be combined with numerous CMOS-IC on the same substrate. (**b**) SEM image of Si microprobe array, each with 160 μm in length and 3.5 μm in diameter at tips. Pictures courtesy Dr. T. Kawano.

Yoon et al. (42) from the Seoul National University reported the process of making a silicon depth-probe microelectrode array by using a combination of plasma and wet etching technique. Neural recordings were obtained from rat somatosensory cortex using a four-site probe. Each electrode had an area of 1,600 μm^2 and they are separated 150 μm from each other. Williams et al. (24) at the Arizona State University described a process for obtaining chronic, multi-site unit recordings using relatively simple and inexpensive implantable multi-channel probe made from different materials and different geometries. The array consists of 33 micro-wires arranged in three parallel rows, made from 35 μm diameter tungsten wires with polyamide insulation giving a total diameter of approximately 50 μm. Recordings were made in the cerebral cortex of awake animals from periods of 3 months or more. They also described a protocol, which demonstrates that these electrodes could provide a model system to study the biocompatibility of neural implants.

5. Conclusion

As described above, there have been many approaches to the construction of sensors for use in neural prostheses or ortheses. They are many characteristics to be achieved such as biocompatibility, duration of implant, functionality, mechanic fixation, selectivity and signal to noise ratios among some of them. Many types of penetrating probes and many different materials and geometries have been tried. However, up to now none of them have proven to be used for clinical chronic recording applications, the reason being that the probes used are insufficiently selective (cuff electrodes), cause damage to the nerve (regenerating electrodes), do not provide information from a large enough volume of the nerve (penetrating electrodes) or their recording qualities diminish after few months of implantation.

Current research includes the study of the geometric properties of different penetrating probes electrodes to achieve, with as few electrodes as possible, and with the simplest signal processing approach, a sensor which can provide a good performance with high selectivity and signal to noise ratios, but across a larger area of the nerve for peripheral nerve recordings. This would be achieved by a novel approach combining multiple-site recordings of the same action potentials with signal processing to record from more distant neurons than those immediately next to the electrode, overcoming the drawback of penetrating probes and allowing measurements to be taken across more of the nerve.

References

1. Hausser, M. (2000). The Hodgkin-Huxley theory of the action potential. *Nature Neuroscience* **3**: 1165.

2. Cole, K. and H. Curtis (1939). Electric Impedance of the squid giant axon during activity. *Journal of General Physiology* **22**: 649–670.

3. Hodgkin, A. and A. Huxley (1939). Action potentials recorded from inside a nerve fibre. *Nature* **144**: 710–712.

4. Hodgkin, A. and B. Katz (1949). *Journal of Physiology* **108**: 37–77.

5. Hodgkin, A. and A. Huxley (1952). A quantitative description of membrane current and its application to conduction and excitation in nerve. *Journal of Physiology* **117**: 500–544.

6. Avery, R. (1973). Implantable nerve stimulating electrode. U.S.A., U.S. Patent #3,774,618.

7. Hagfors, N. (1972). Implantable electrode. U.S.A., U.S. Patent #3,654,933.

8. Popovic, D., R. Stein, K. Jovanovic, R. Dai, A. Kostov and W. Armstroing (1993). Sensory nerve recordin for closed-loop control to restore motor functions. *IEEE Transactions on Biomedical Engineering* **40**(10): 1024–1031.

9. Naples, G., J. Sweeney and J. Mortimer (1986). Implantable cuff, method of manufacture, and method of installation. United States Patent. USA, No. 4,602,624.

10. Klepinski, R. (1994). Implantable neural electrode. USA, #5,282,486.

11. Sahin, M. and D. Durand (1998). Improved nerve cuff electrode recordings with subthreshold anodic currents. *IEEE Transactions on Biomedical Engineering* **45**(8): 1044–1050.

12. Takeuchi, S. and I. Shimoyama (1999). Wireless recording of insect neural activity with an SMA microelectrode. *Proceedings of The First Joint BMES/EMBS Conference Serving Humanity, Advancing Technology*, Atlanta, USA, IEEE.

13. Jezernik, S. and T. Sinkjaer (1999). On statistical properties of whole nerve cuff recording. *IEEE Transactions on Biomedical Engineering* **46**(10): 1240–1245.

14. Nakatani, H., T. Watanbe and N. Hoshimiya (2001). Detection of nerve action potentials under low signal-to-noise ratio condition. *IEEE Transactions on Biomedical Engineering* **48**(8): 845–849.

15. Jensen, W., S. Lawrence, R. Riso and T. Sinkjaer (2001). Effect of initial joint position on nerve-cuff recordings of muscle afferents in rabbits. *IEEE Transactions on Neural Systems and Rehabilitation Engineering* **9**(3): 265–273.

16. Hoffer, J. and K. Kallesoe (2001). How to use nerve cuffs to stimulate, record or modulate neural activity. In: *Neural Prostheses for Restoration of Sensory and Motor Function*. J. K. Chapin (ed.), CRC Press, 139–175.

17. Edell, D., J. Churchill and I. Gourley (1982). Biocompatibility of a silicon based peripheral nerve electrode. *Biomaterials, Medical Devices, and Artificial*. 103–122.

18. Akin, T., K. Najafi, R. Smoke and R. Bradley (1994). A micromachined silicon sieve electrode for nerve regeneration applications. *IEEE Transactions on Biomedical Engineering* **41**(4): 305–315.

19. Kovacs, G. (1990). *Technology Development for a Chronic Neural Interface. Electrical Engineering*, Stanford University.

20. Mannard, A., R. Stein and D. Charles (1974). Regeneration electrode units: Implants for recording from single peripheral nerve fibers in freely moving animals. *Science* **183**: 547–549.

21. Kovacs, G., C. Storment, et al. (1994). Silicon-substrate microelectrode arrays for parallel recording of neural activity in peripheral and cranial nerves. *IEEE Transactions on Biomedical Engineering* **41**(6).

22. Wallman, L., Y. Zhang, T. Laurell and N. Danielsen (2001). The geometric design of micromachined silicon sieve electrodes influences functional nerve regeneraion. *Biomaterials* **22**: 1187–1193.

23. Stieglitz, T., H. Ruf, M. Gross, M. Schuettler and U. Meyer (2002). A biohybrid system to inteface peripheral nerves after traumatic lesions: design of a high channel sieve electrode. *Biosensors and Bioelectronics*, 1–12.

24. Williams, J., R. Rennaker and D. Kipke (1999). Long-term neural recording characteristics of wire microelectrode arrays implanted in cerebral cortex. *Brain Research Protocols* **4**: 303–313.

25. Grundfest, H., R. Sengstaken and W. Oettinger (1950). Stainless steel micro-needle electrodes made by electrolytic pointinc. *Physical Instruments for the Biologist* **21**(4): 360–361.

26. Kruger, J. (1983). Simultaneous individual recordings from many cerebral neurons: techniques and results. *Reviews of Physiology, Biochemistry & Pharmacology* **98**: 177–233.

27. Gross, G., E. Rieske, G. Kreutzberg and A. Meyer (1977). A new fixed-array multi-microelectrode system designed for long-term monitoring of extracellular single unit neuronal activity in vitro. *Neuroscience Letters* **6**: 101–106.

28. Pine, J. (1980). Recording action potentials from cultured neurons with extracellular microcircuit electrodes. *Journal of Neuroscience Methods* **2**: 19–32.

29. Pickard, R., A. Collins, P. Joseph and R. Hicks (1979). A flexible printed circuit probe for electrophysiology. *Medical and Biological Engineering Computing* **17**: 261–267.

30. Wise, K., J. Angell and A. Starr (1970). An integrated-circuit approach to extracellular microelectrodes. *IEEE Transaction on Biomedical Engineering* **BME-17**(3): 238, 247.

31. Wise, K. and J. Angell (1975). A low-capacitance multielectrode probe for use in extracellular neurophysiology. *IEEE Transactions on Biomedical Enginnering* **BME-22**(3): 212, 219.

32. Prohaska, O., F. Olcaytug, K. Womastek and H. Petsche (1977). A multielectrode for intracortical recordings produced by thin-film technology. *Electroencephalography and Clinical Neurophysiology* **42**: 421–422.

33. Ensell, G., Banks, D.J., Richards, P.R., Balachandran, W. and Ewins, D.J. (2000). Silicon based microelectroes for neurophysiliogy, micromachined from silicon-on-insulator wafers. *Medical and Biological Engineering Computing* **38** 175–179.

34. Bai, Q. and K. Wise (2001). Single-unit neural recording with active microelectrode array. *IEEE Transactions on Biomedical Engineering* **48**(8): 911–920.

35. Xu, C., W. Lemon and C. Liu (2002). Design and fabrication of a high-density metal microelectrode array for neural recording. *Sensor and Actuators A* **96**: 78–85.

36. McNaughton, T. and K. Horch (1996). Metallized polymer fibers as leadwires and intrafascicular microelectrodes. *Journal of Neuroscience Methods* **70**: 103–110.

37. Nordhausen, C., E. Maynard and R. Normann (1996). Single unit recording capabilites of a 100 microelectrode array. *Brain Research* **726**: 129–140.

38. Maynard, E., T. Nordhausen and R. Normann (1997). The Utah Intracortical Electrode Array: A recording structure for potential brain-computer interfaces. *Electroencephalography and Clinical Neurophysiology* **102**: 228–239.

39. Rousche, P. and R. Normann (1998). Chronic recording capability of the Utah Intracortical Electrode Array in cat sensory cortex. *Journal of Neuroscience Methods* **82**(1): 1–15.

40. Stieglitz, T. and M. Gross (2002). Flexible BIOMEMS with electrode arrangements on front and back side as key component in neural prostheses and biohybrid systems. *Sensors and Actuators B* **83**: 8–14.

41. Kawano, T., Y. Kato, M. Futagawa, H. Takao, K. Sawada and M. Ishida (2002). Fabrication and properties of ultrasmall Si wire arrays with circuits by vapor-liquid-solid growth. *Sensor and Actuators A*: 1–7.

42. Yoon, T., E. Hwang, D. Shin, S. Park, S. Oh, S. Jung, H. Shin and S. Kim (2002). A micromachined silicon depth probe for multichannel neural recording. *IEEE Transactions on Biomedical Engineering* **47**(8): 1082–1087.

43. Kim, J. H., Manuelidis, E. E., Glenn, W. W., Fukuda, Y., Cole, D. S. and Hogan, J. F. (1983) Light and electron microscopic studies of phrenic nerves after long-term electrical stimulation. *Journal of Neurosurgery* **58**, 84–91.

44. Heiduschka, P. and S. Tanos (1998). Implantable bioelectronic interfaces for lost nerve functions. *Progress in Neurobiology* **55**(5): 433–461.

45. Ferris, C. (1974). *Introduction to Bioelectrodes*, Plenum Press.

Chapter 7

Impedance Spectroscopy and Optical Analysis of Single Biological Cells and Organisms in Microsystems

Shady Gawad, David Holmes, Giuseppe Benazzi, Philippe Renaud, and Hywel Morgan

Abstract

A novel microfabricated flow cytometer for simultaneous impedance and optical measurement of single cells and particles is described in this chapter. We discuss the sensitivity of the system with regard to the impedance sensor and describe the optical setup. The relevant parameters related to the experimental setup and sample preparation are discussed. The use of dielectrophoretic forces for particle manipulation is presented as a simple enabling technology, which allows the manipulation of particles within microfluidic devices. The fabrication processes required to produce the impedance sensor chips are described with relevance to the chip design and features. Finally, the system is used to discriminate between different marine algae populations.

Key words: Impedance spectroscopy, dielectrophoresis, electrorotation, dielectric, single cell, algae, microfluidics, microfabrication.

1. Introduction

For the last 10 years, a number of research efforts aimed at health sciences and biological and chemical applications have looked into system miniaturization to provide faster, smaller, and more efficient solutions for cell and tissue analysis and manipulation. The principal areas of interest include drug discovery and delivery, tissue engineering, surgical instruments, and diagnostic tools. This work has shown that interfacing biomaterials with microstructured silicon or glass devices has wide utility and is gaining momentum as more and more successes are reported in the literature.

As of 2006, over a hundred groups worldwide are active in the field of BioMEMS and micro Total Analysis Systems (μTAS). This last acronym is synonymous of lab-on-a-chip and covers

M.P. Hughes, K.F. Hoettges (eds.), *Microengineering in Biotechnology*, Methods in Molecular Biology 583,
DOI 10.1007/978-1-60327-106-6_7, © Humana Press, a part of Springer Science+Business Media, LLC 2010

microfabricated integrated systems for preparation, separation, reaction, detection, and synthesis in chemistry and biology. Different approaches are typically observed in µTAS, either based on microarrays or microfluidic systems, to increase data throughput and resolve smaller signals.

In the field of flow cytometry, optical techniques have been subject to an impressive development in the last 20 years due to the availability of lasers and fluorescence markers and the invention of the fluorescence-activated cell sorter (FACS). Despite this, little work has been carried out to miniaturize these expensive and complex room-filling laboratory instruments, which are still only found in a relatively small number of clinical and scientific research centers.

Until recently, advances made in the development of the electrical complement of the FACS, the Coulter counter, were quite limited and the data acquired with standard Coulter equipment essentially focuses on cell sizing, viability, as well as sample concentration.

Publications on both miniaturized optical and electrical cytometers have appeared in the last few years. Both approaches, optical and electrical, present comparative advantages and drawbacks. It is generally argued that the fluorescent based, or optical approach, is more sensitive and recognized cell discrimination technique, but that it comes at the cost of cell tagging and thus cell modification. In contrast, electric impedance techniques are label-free which simplifies sample preparation and allows applications requiring unmodified cell retrieval, for instance, when cell tagging would trigger specific cell response paths.

In both cases, advantages are offered by the integration and miniaturization of these devices; allowing the possibility to work on single cells from a small sample and even to consider each cell as a unique sample. In the foreseeable future integrated systems will allow a number of operations to be performed automatically such as cell sampling, preparation, culture, and detection or monitoring of cellular reaction to drugs.

For diagnostic applications, a miniaturized cytometer will considerably reduce the related lab costs and increase efficiency. Eventually, the small size of such devices will make their use in point-of-care applications possible. For drug research, reduction in the amount of reagent or drug required to perform a specific test translates into faster and cheaper screening.

2. Cell Level Discrimination

To characterize and distinguish different cell types or cell states, a number of techniques have been developed. These either consider the cell in its entirety (e.g., size, shape, granularity, and light absorbance) or aim at specific elements on the cell surface or within the cytoplasm. Some detection techniques measure an intrinsic physical or physiological characteristic (e.g., electric, optic, deformability, and density) of the cell or its sub-components, while others rely on markers (e.g., fluorescent and magnetic) which highlight a specific constituent. Markers allow very accurate detection down to the molecular level, but they generally require a preliminary incubation procedure and by the very nature of this may alter the cell. This is especially true for cells from the immune system, where the cell's sensitivity to various antigens may be influenced in different ways by antibody to CDs (1, 2).

The time necessary for a tagging molecule to meet and bind to its target is limited by the concentration, diffusion properties and position or accessibility (i.e., due the presence of physical barriers) of these molecules. Using micrometer size capillary channels or if necessary integrated mixers the diffusion time can be reduced significantly in small chambers providing a reaction in a fraction of the time necessary within a standard test tube. An additional problem is that some non-surface markers need to enter the cell in order to reach their specific target. This often requires a membrane permeabilization step and in many cases this practice is thought to affect further analysis performed on the cell.

A number of optical detection based microsystems have been shown in the literature, both for analysis and cell sorting applications (3–5). These systems show much promise and with further development will lead to commercial devices in the next few years.

For some applications fluorescent labeling is not necessary, for example, where the intrinsic optical properties of the particles under analysis are such that they exhibit autofluorescence. The study of phytoplanktonic algae is one such area, with the different algal taxa contain different photosynthetic pigments. Measurement of the fluorescence emitted by the algae upon excitation with different laser wavelengths can be used to identify species as will be shown in the measurement section of this chapter. The application of optical flow cytometry to single-cell microbial populations analysis has been reviewed by Davey and Kell (6).

Miniaturized electrical approaches are generally based on an AC electrokinetic or impedance spectroscopy technique (7), which will be discussed in detail in this chapter. Both approaches are marker-free and particularly suited to operate at the single-cell level. Due to differences in their complex dielectric properties and dimensions, specific information from cellular sub-structures such as the cell membrane, cytoplasm, or nucleus can be obtained by measurement at distinct frequencies.

AC electrokinetic forces (8) in dielectrophoresis (DEP) (9, 10) or torque in electrorotation (ROT) (11–13) can be applied to single or groups of cells for manipulation in microsystems. This can done in combination with an additional pressure driven flow to sort particles according to their dielectric properties (10–14).

Dielectrophoretic separation can be achieved by selecting a frequency where the amplitude or direction of the applied forces are sufficiently large to push, hold, or deflect a specific cell type while leaving other cell types unaffected. Electrorotation measurements of cell properties are generally done by video analysis of the rotational speed of a cell subjected to a rotational electric field over a range of frequencies. AC electrokinetic experiments are generally done in low conductivity saline in order to enhance the range of forces that can be applied and reduce Joule heating effects. The applied electric field is generally less than $10^4\,\mathrm{V\,m^{-1}}$ but can reach up to $10^6\,\mathrm{V\,m^{-1}}$ at the MHz frequency range, without noticeably damaging the cells. In the lower frequency range, where the cell membrane will polarize more easily, electro-permeabilization treatment or electro-bursting of cells can occur for electric field strengths around $10^5\,\mathrm{V\,m^{-1}}$. For more information on background theory in the field of AC electrokinetics several texts are available (15–16).

In traditional cell dielectric spectroscopy electrical properties of a cell suspension are measured using volumes sometimes as small as 0.1 ml (17–19). To obtain a reasonable signal and limit particle–particle interaction, the volume fraction of particles in suspension is typically in the range of 1–30%. Still, even in such cases where a small volume and small volume fractions are used, the resulting measurement gives an average over a large number of cells. This represents a significant drawback of such systems as the discrimination of cell sub-population and cell sorting is impossible. To address these issues, microsystems can be used in order to define microscopic detection zones by defining a pattern of microelectrodes within the microchannels. Cells can then be addressed individually in a sequential manner rather than in bulk. The parameters primarily accessed using this technique are the cell volume, cell membrane capacitance, and conductance, as well as electrical parameters related to the cytoplasm.

3. System Platform and Integration

The integration on a microsystem platform of several complementary tools is a challenging goal, which opens up a realm of possible applications for this technology (**Fig. 7.1**). Optical and electrical means for cell manipulation and analysis have been demonstrated in microfluidic devices by a number of groups.

It has only been in the last couple of years that the technology has matured sufficiently to allow the development of robust analysis systems. The inclusion of new techniques, such as cell bursting and single-cell genetic modification, achievable in integrated

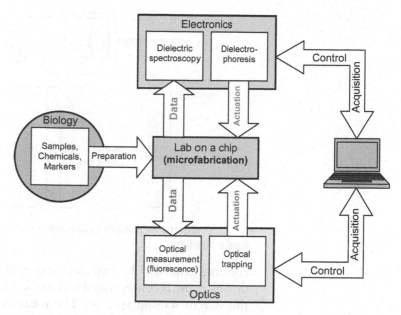

Fig. 7.1. Schematic of a Lab-on-a-chip system comprising optical and electrical interfaces for cell preparation, manipulation and analysis.

systems by the use of electro-mediated lysis of cells and electro-permeabilization, is opening up the field and will lead to a plethora of novel diagnostic devices in the coming years.

Although progress has been made in developing these miniaturized cell manipulation and diagnostic systems, it is still the case that samples are generally processed off-chip using lengthy, labour-intensive preparation methods. Proper sample preparation is especially important, as certain stages of a microsystem may have low tolerance to large debris, which may cause obstruction of the system's microfluidics and result in failure of device operation. Filters and cleaning techniques have been proposed to alleviate part of the problem; still, achieving good results relies heavily on the initial sample preparation.

Laboratory automation and computer control play equally important roles. Results beyond the proof of concept stage require a good degree of repeatability, both in processing samples and monitoring controlling parameters. Automation in processing the acquired data is also necessary and permits one to rapidly detect any problem with the system operation and apply corrective steps.

4. Detection Techniques

4.1. Impedance Measurement of Single Cells

The Coulter method of sizing and counting particles is based on measurable changes in electrical resistance produced by cells or particles passing through a small aperture (**Fig. 7.2**) (20). The

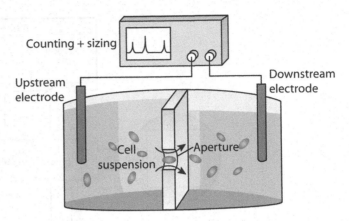

Fig. 7.2. Coulter counter principle. Cells drawn through a small aperture are counted and sized sequentially.

current, generated by two large electrodes placed upstream and downstream, is concentrated within the aperture, which represents the system sensing volume. The measured voltage pulse height produced as the particle passes through the aperture is proportional to the volume of the particle. Several thousand particles per second can be individually counted and sized. In addition, a metering device is used to draw a known volume of the suspension through the aperture; a simple count of the number of pulses can then yield the concentration of particles in the sample. In applications such as peripheral blood differential leukocyte count, statistic sizing is generally sufficient to detect different types of leukocytosis or leukopenia. Forcing both the electrical current and the cell stream into a small aperture is a simple method to generate a highly sensitive detection volume to achieve single-cell detection. As a first approximation, the sensitivity of the measurement increases with the selection of a smaller hole diameter. The limit is that the aperture has to be big enough to accommodate the passage of the largest particle in the suspension.

4.1.1. On-Chip Examples

The first μCoulter devices were proposed by Larsen et al. (21) and Koch et al. (22, 23). These systems were microfabricated in silicon and use microfluidic sheath flows and filter structures. Reports of on-chip measurement were published later by Fuller et al. on granulocytes (24), Larsen et al. on somatic cells using a simple silicon aperture (25) and Gawad et al. on erythrocytes using a coplanar electrode geometry (26). The original μCoulter technique is clearly limited to cell sizing by the use of a single measurement frequency, generally at low frequency or DC. Measurement at multiple frequencies is necessary to determine other attributes of the cell, similar to what is obtained in traditional dielectric spectroscopy. An integrated cytometer device for stop-cell measurements using a frequency sweep measurement technique was presented by Ayliffe et al. using electroplated gold electrodes (7).

*4.1.2. Electric Field
Geometry and
Microelectrodes*

The electric field geometry is a very important aspect of the system as it is directly related to the shape and amplitude of the detected impedance signal. According to Deblois, the variation of the resistance ΔR due to a small particle of diameter d passing in an aperture of diameter D is given in first approximation as

$$\Delta R = \frac{4\rho d^3}{\pi D^4} \quad \text{as } D^4 \propto J^{-2} \quad \text{then} \quad \Delta R \propto \rho d^3 J^2, \qquad [1]$$

where J is the current density in the tube section. Considering this physical aperture as a current tube, it can be seen that the amount of information gained from a specific location by impedance measurement using a bipolar electrode configuration is proportional to the square of the local current density.

Consequently, some of the important differences that exist between on-chip cell impedance analysis and the traditional Coulter instrument are related to the use of microelectrodes and micro fluidic geometries to define the shape of the electric field geometry. As mentioned above, in the Coulter example, the measuring electric field is produced by relatively large electrodes on each side of a small aperture. This approach offers some significant advantage at low frequency due to the increased electrode surface area and reduced impedance of the electrode/electrolyte interface. By reducing the device dimensions to fit the system in a microfluidic chip, two possibilities or levels of integrations may be considered:

(a) The first solution is to keep the size of the electrodes relatively large ($200 \times 200 \ \mu m^2$ range) and use a microfabricated aperture or a constriction in the channel section. This approach requires careful analysis of the inhomogeneous distribution of the current density at the electrode surface. This effect, especially in the case of shallow channels can significantly reduce the electrode effective area. In the case of impedance spectroscopy at frequencies up to 20 MHz, large electrodes can also lead to higher stray capacitances and limit the sensitivity for high frequency measurements. Integration of other systems such as cell tracking and sorting with practical cell dilutions ($10^6 \ ml^{-1}$) will be more difficult due to the larger volumes of liquid in the system. In a sandwich structure, electrodes patterned only on the bottom side of the channel may be used. In addition, a certain distance must be set between the electrodes and the constriction or aperture of approximately three to five times the channel height; this is to establish a homogeneous electric field in the vertical direction.

(b) Alternatively, it may be decided to further reduce the size of the electrodes down to the characteristic dimension of the cells or particles under study which has several implications.

In this case, the detection area is directly defined by the electrode geometry within the capillary channel. This approach reduces stray capacitances and thus increases the device operating bandwidth toward higher frequencies. Measurements at the higher end of the β-dispersion (see below) are then possible. Small channel cross-sections permit sequential tracking of cells along the channel and the integration of features such as sorting by partitioning the channel length for successive operations. Small microelectrodes result in an apparent dilemma; on the one hand, a higher sensitivity results from the higher field line concentration around the cell. On the other hand, the high interface impedance (due to the electrical double layer) makes measurement at low frequency more difficult. To alleviate this problem there are some techniques to increase the effective surface area of microelectrodes based on nanoporous films such as platinum black, iridium oxide, or titanium nitride (27, 28).

Nanoporous films can be deposited on a seed electrode through an additional fabrication step. For platinum black, which uses electro-deposition, the effective surface area increase can reach 100- to 1,000-fold, but in practice it is particularly difficult to obtain a high reproducibility of the final interface impedance. Alternatively, the use of a sputtered iridium oxide film (IrOx) demonstrates reproducible results and large effective surface area increase following activation (29). The activation is simply performed by applying a cyclic voltametry step of 10–20 cycles of 3 V at 0.1 Hz to the microelectrodes with the microfluidic channel filled with NaCl. It can be seen in **Fig. 7.3** that after activation the low frequency cut-off point (taken at a 45° phase) is decreased from ~100 kHz down to the 1 kHz range. The electrodes physical dimensions are in this case 20 × 40 μm.

Another issue arises from the inhomogeneous current density, be it either within the channel constriction (i.e., aperture) or if small electrodes are used between the closely placed electrode pair.

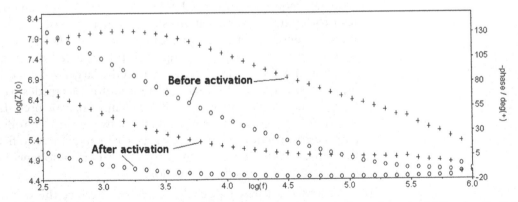

Fig. 7.3. Frequency response analysis showing impedance amplitude and phase of a pair of IrOx microelectrodes as measured before and after the activation procedure.

This inhomogeneity can bias the detection signal peak, as a function of the particle's trajectory, as it passes through the measurement volume. In traditional Coulter systems, the length and shape of the aperture plays an important role in controlling the uniformity of the signal shape and this has been extensively studied (30). Similar studies may be performed for the integrated chip approach with embedded microelectrodes. A complex dielectric model is used to determine the change in electric current between a pair of electrodes due to passage of a biological cell. The influence on the dielectric spectrum of different cell properties, such as membrane capacitance and cytoplasm conductivity is of particular interest. The measurement volume is located between a pair of facing microelectrodes in a microchannel filled with a saline solution. A 3D finite element model is used to determine the electric field in the channel and the resultant changes in charge densities at the measurement electrode boundaries as a cell flows through the measurement volume. The charge density is integrated on the electrode surface to compute the displacement current and the channel impedance for a given applied alternating electrode voltage and frequency. The excitation signals are taken to be of suitably low frequency and low applied voltage, allowing the quasi-static approximation and linear behavior of the system to be assumed. In the present example, the measurement volume geometry details are for the specific case of a sensor with a channel of $20 \times 20 \ \mu m^2$ cross-section and two facing electrodes, $20 \times 20 \ \mu m^2$ in area, on opposite walls of the channel (**Fig. 7.4**). These channel dimensions are chosen in order to determine the sensitivity of a sensor, which could accommodate cell diameters up to 15 μm without clogging the channel.

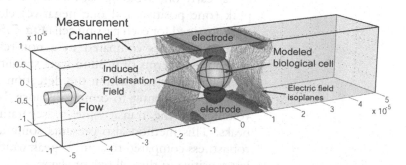

Fig. 7.4. Diagram of flow cytometry measurement in a microchannel with integrated microelectrodes. The modeled cell passes between the electrode pair, altering the electric field distribution for different excitation frequencies, thus giving rise to a change in the measured channel impedance spectrum. The electric field density in the channel represented here by isosurfaces shows typical polarization effects of the Maxwell-Wagner type (caps on the top and bottom of the modeled cell) and the inhomogeneous field due to the electrode pair. (From Gawad et al. (31) – Reproduced by permission of The Royal Society of Chemistry)

A dielectric sphere of equivalent complex permittivity is used as a simplified model to describe a biological cell. The model includes various geometrical parameters of the cell, such as size and position in the channel. Using such a model it is shown that given a reasonable size and placement of the cell along the center line of the channel, a good signal reproducibility can be obtained. Also using analytical methods, such as conformal mapping, compensation factors can be found, depending on the sensor geometry, to account for the fringing effect and non-uniformity of the electric field. Numerical approaches have been used to determine the practical limits of this compensation method (31).

4.1.3. Cell Detection Area and Differential Measurement

In practice, the measurement area of the chips contains two measurement volumes (pairs of electrodes) in close proximity. In the measurement area the channel width is reduced in order to decrease the detection volume and thus increase the electrical sensitivity of the system to the passage of the cell. This restriction also decreases the chance of having two cells entering the measurement area simultaneously. The electrode areas are typically $20 \times 40 \ \mu m^2$ and channel cross-section $20 \times 40 \ \mu m^2$. The distance between the two electrode pairs is 60 μm.

The particle or cell moving under pressure-driven flow successively pass through the two electric field regions, thus consecutively modifying the current through each detection volume. The use of a differential measurement scheme significantly reduces the measurement noise; small thermal fluctuations or variations in the composition of the suspending fluid are thus cancelled out. The chip is connected to a circuit which uses differential electronics to amplify the small difference of current passing through the two sensors. This approach allows rejection of the common mode signal early on, avoiding saturation of the amplifiers. A double peak (one positive and one negative) electrical signal shape is thus obtained for each passing cell (**Fig. 7.5**). This approach offers many advantages compared to a non-referenced measurement technique in terms of ground level stability and signal amplification. The cell speed within the detection channel is obtained by dividing the center-to-center distance between the detection volumes by the transit time t_{tr} separating the two measured peaks. This method also presents some advantages in terms of robustness compared to a single peak width measurement, as it is less sensitive to the cell size or shape.

4.1.4. Further Techniques

4.1.4.1. Guard Ring

The addition of a guard ring in order to straighten the electric field lines around the current sensing electrodes is a practical approach that addresses the issue of the electric field uniformity and permits simplification of the calculation of the cell geometric factor. The use of grounded guard electrodes in conjunction with current sensing electrodes connected to a virtual ground in auto-balanced

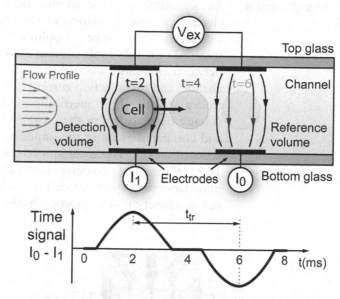

Fig. 7.5. Cell detection principle and expected electrical differential current signal for a passing cell. The detection and reference volumes switch as the cell passes through each. The cell speed in the measurement channel is computed by measuring the transit time t_{tr} separating the two peaks. The bottom electrodes are kept at a virtual ground potential.

bridge configuration is preferred (**Fig. 7.6**). This is because at higher frequencies (MHz and greater) it is difficult to implement non-phase lagging followers to drive active guard electrodes, as would be required by a traditional bridge electrical circuits. This amplification approach also reduces the effect of the stray capacitances as there is no current leakage from the current sensing leads, at a virtual ground potential, to the ground shielding or guard electrodes leads.

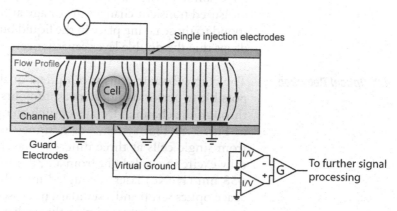

Fig. 7.6. Principle and geometrical approach to implement guard electrodes in a differential auto-balanced amplification scheme.

4.1.4.2. 4 Point Differential Electrodes

An alternative method to that of nanoporous modification of electrodes surfaces, to increase the effective charge density of the electrodes, is the use of a 4-point differential electrode configuration (32). In this configuration an alternating voltage is applied between two large microelectrodes placed at either end of the channel and the injection current is measured. Three pickup electrodes are placed symmetrically in the channel. The differential voltage drops between the pickup electrodes are then amplified and compared yielding the impedance variation due to the presence of the cell. This technique has been shown to work well up to 10 kHz using the geometry shown in **Fig. 7.7a**. In **Fig. 7.7b** the impedance variation measured at 1 kHz is represented for a passing cell as a function of its position in the channel.

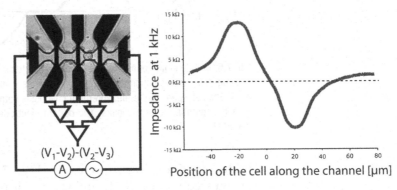

Fig. 7.7. (**a**) 4-point differential measurement principle. (**b**) Measured impedance signal. (Figure courtesy of T. Braschler)

The pickup electrodes surface capacitance, due to the double layer, still subsists in the tetrapolar measurement scheme. But since the input impedances used in the amplification is high only negligible amounts of current flow through those electrodes. The measured transient changes in voltage at the electrode thus correspond to those taking place in the liquid bulk with minimal voltage drops due the double layer capacitance.

4.2. Optical Detection

The chip substrate consist of optical quality glass, it is therefore possible to perform optical measurements on the cells as they pass through the channel. Details of such an optical system are shown in **Fig. 7.8**. This system was designed to measure fluorescence from single cells at three different wavelength ranges. To optically excite the cells light from a 532 nm (solid state YAG) and a 633 nm (HeNe) laser is coupled into the channel using a free-space optics setup and a standard microscope objective. The laser beams are combined and pass through a beam expander into an infinity corrected objective lens ($\times 20$, 0.5 N.A. Nikon). For

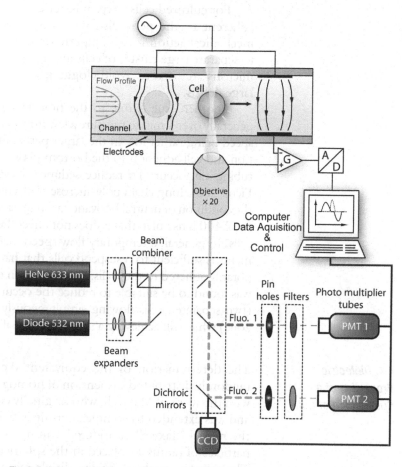

Fig. 7.8. Schematic of the single-cell optical detection system using two excitation lasers and up to three emission detectors.

detection the fluorescent light emitted from the cells is collected by the same objective lens is spectrally and spatially filtered by passing through dichroic mirrors, band pass filters, and lens and pinholes positioned in front of photomultiplier tubes (Hamamatsu). The intensity of the light at the focal point of the objective (optical detection zone in the channel) is typically 600 μW. The emission filters have center wavelengths of 585 nm and 675 nm and a longpass filter with cutoff at 710 nm (Chroma, Rockingham, VT, USA).

4.3. Sample Preparation and Chip Maintenance

Sample preparation is certainly one of the most time consuming and critical aspect of the system operation. The quality of the sourced sample is important as is proper lab technique for cell harvesting, filtering, labeling, and washing. The large amount of cell debris, DNA fragments or other substances, which accumulates on the channel walls necessitates regular cleaning of the microchannel.

For cultured cells, trypsin is generally used to re-suspend cells adherent to the petri dish wall. For cells grown in suspension mechanical action using a pipette or gentle vibration is sufficient to separate large cluster of cells into single units. Mesh filters of 50 microns are used to avoid clogging of the capillary channel with large debris.

Another issue is due to the time taken for the measurement, generally over 1 min to measure a few hundreds of cells. With the flow speed being rather low in the larger parts of the channel sedimentation and cell adhesion to the bottom glass surface of the channel or tubing can occur. To reduce sedimentation a small percentage of Ficoll 400 a long chain poly-sucrose molecule can be used to obtain the condition of neutral buoyancy as suggested by Holmes et al. (33). Ficoll400 is useful in that is does not effect the osmotic balance of the cells. In general, a temporary flow speed increase between measurements is sufficient to resuspend cells that have sedimented and temporarily adhered to the walls. A cell dilution of up to $\sim 10^6$ cells ml^{-1} was found to be suitable to reduce the occurrence of doublet events (two separated cells flowing simultaneously through the detection zone) while still achieving a steady stream of cells.

4.4. Dielectric Properties of a Suspension of Cells

The determination of the equivalent dielectric properties of a randomly distributed suspension of homogeneous particles started with the work of Maxwell, who originally considered conductivity, and was extended to complex permittivity by Wagner. In deriving the model Maxwell considered a number N_p of non-interacting particles of radius R, placed in the spherical volume of radius R_0. The sum p_{sum} of their effective dipole moment p_{eff} contributions is simply obtained by

$$p_{\text{sum}} = N_p p_{\text{eff}} \qquad [2]$$

Considering a sphere filled with a homogeneous material with a complex permittivity $\tilde{\varepsilon}_2$, suspended in a liquid of complex permittivity $\tilde{\varepsilon}_1$, the effective dipole moment of the sphere p_{sphere} is then given by

$$p_{\text{sphere}} = 4\pi R_0 \varepsilon_1 \left(\frac{\tilde{\varepsilon}_2 - \tilde{\varepsilon}_1}{\tilde{\varepsilon}_2 + 2\tilde{\varepsilon}_1} \right) \qquad [3]$$

A complex permittivity $\tilde{\varepsilon}$ is defined as

$$\tilde{\varepsilon} = \varepsilon_r - \frac{i\sigma}{\omega \varepsilon_v} \qquad [4]$$

σ is the conductivity of the material, ε_r is the relative permittivity, ε_v the permittivity of free space, and ω the angular frequency.

By equating the relations [2] and [3], the complex form of Maxwell's expression gives an expression for the equivalent complex permittivity of the suspension $\tilde{\varepsilon}_{\text{equ}}$.

$$\frac{\tilde{\varepsilon}_{equ} - \tilde{\varepsilon}_1}{\tilde{\varepsilon}_{equ} + 2\tilde{\varepsilon}_1} = \phi \frac{\tilde{\varepsilon}_2 - \tilde{\varepsilon}_1}{\tilde{\varepsilon}_2 + 2\tilde{\varepsilon}_1} \qquad [5]$$

The volume fraction φ is defined by $\phi = N_p \left(R^3 / R_0^3 \right)$. An elegant formulation of the equivalent complex permittivity $\tilde{\underline{\varepsilon}}$ is given by Jones[15]

$$\tilde{\varepsilon}_{equ} = \tilde{\varepsilon}_1 \frac{1 + 2\phi \tilde{K}(\tilde{\varepsilon}_2, \tilde{\varepsilon}_1)}{1 - \phi \tilde{K}(\tilde{\varepsilon}_2, \tilde{\varepsilon}_1)} \qquad [6]$$

where $\tilde{K}$ is the complex Clausius–Mossotti factor which contains the magnitude and phase information of the dipole representing the cell defined as

$$\tilde{K} = \frac{\tilde{\varepsilon}_2 - \tilde{\varepsilon}_1}{\tilde{\varepsilon}_2 + 2\tilde{\varepsilon}_1} \qquad [7]$$

An expression for more complex particle models including multiple shells is obtained by substituting the corresponding Clausius–Mossotti factor to account for additional relaxations in the dielectric properties of the cell (**Fig. 7.9**).

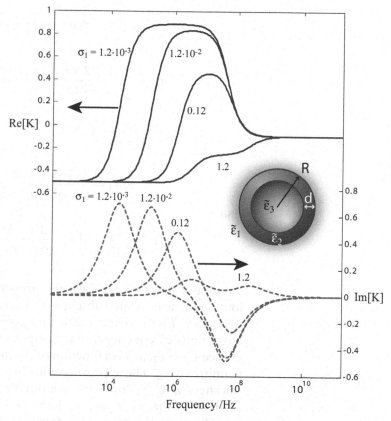

Fig. 7.9. Frequency dependence of the real and imaginary part of the Clausius–Mossotti factor for a shell-covered sphere in suspending media of different conductivity, with parameters $\sigma_1 = 1.2 \cdot 10^{-3} - 1.2\, \mathrm{S \cdot m^{-1}}$, $\sigma_2 = 1 \cdot 10^{-8}\, \mathrm{S \cdot m^{-1}}$, $\sigma_3 = 0.5\, \mathrm{S \cdot m^{-1}}$, $\varepsilon_1 = 78 \cdot \varepsilon_v$, $\varepsilon_2 = 10 \cdot \varepsilon_v$, $\varepsilon_3 = 60 \cdot \varepsilon_v$, $R = 2\, \mu m$, $d = 10\, nm$.

The number of theoretical relaxations (drops in permittivity with increasing frequency) is equal to the number of interfaces in the cell model. In particular, the single-shell model displays two relaxations: one for the interface between the external medium and the membrane and the second for the membrane internal interface with the cytoplasm.

Even in the simple case of a sphere covered by a single shell, the expression can get rather intricate and can become impractical for studying experimental data. Pauly and Schwan were the first to publish a convenient set of equations detailing the dispersion parameters for such a model (34). Although the single-shell model theoretically gives two separate relaxations when considering the full Pauly and Schwan expressions, permittivity measurements in physiological saline solution show only one relaxation in practice. Its characteristic frequency is located in the MHz range and is related to the time needed to charge the membrane (subscript 2) through the intra-(subscript 3) and extracellular (subscript 1) media. The simplified expressions describing this dispersion are

$$\sigma_s = \sigma_1 \frac{(1-\phi)}{1+\phi/2}, \qquad [8]$$

$$\sigma_\infty = \sigma_1 \frac{1 + 2\phi(\sigma_3 - \sigma_1)/(\sigma_3 + 2\sigma_1)}{1 - \phi(\sigma_3 - \sigma_1)/(\sigma_3 + 2\sigma_1)}, \qquad [9]$$

$$\varepsilon_s = \varepsilon_1 \frac{(1-\phi/2)}{(1+\phi)} + \Delta\varepsilon \text{ with } \Delta\varepsilon = \frac{9}{\varepsilon_v} \frac{\phi}{(2+\phi)^2} R\, C_2 \quad [10]$$

$$\varepsilon_\infty = \varepsilon_1 \frac{(1+2\phi)\varepsilon_3 + 2(1-\phi)\varepsilon_1}{(1-\phi)\varepsilon_3 + (2+\phi)\varepsilon_1}, \qquad [11]$$

and

$$T = \frac{1}{2\pi f_c} = R\, C_2 \left(\frac{1}{\sigma_3} + \frac{1}{\sigma_1} \frac{(1-\phi)}{(2+\phi)} \right). \qquad [12]$$

Subscript s relates to the dielectric properties of the suspension at low frequencies while subscript ∞ relates to the values at high frequency. T is the characteristic time constant and is related to the characteristic frequency f_c of the dispersion. The membrane capacitance $C_2 = \varepsilon_2 \varepsilon_0 / d$ is a function of the membrane thickness d and permittivity ε_2. The approximations involved here are small shell thickness $d \ll R$, low shell conductivity $\sigma_2 \ll \sigma_1, \sigma_3$, and the assumption that $\sigma_1 \gg \omega \varepsilon_1 \varepsilon_0$ and $\sigma_3 \gg \omega \varepsilon_3 \varepsilon_0$.

The second dispersion is observed with a suspending medium of low conductivity at a higher frequency range. The membrane capacitance is considered short-circuited and the observed relaxation is due to the Maxwell-Wagner dispersion resulting from the difference in permittivity between the relatively conductive cytoplasm and the insulating suspending solution. The permittivity and conductivity of a suspension of shell-covered particles is plotted in **Fig. 7.10** for suspending media of different conductivities.

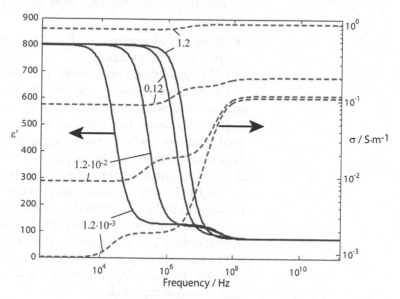

Fig. 7.10. Permittivity and conductivity spectrum for a shell covered sphere in different suspending media conductivities, with $\varphi = 0.2$, $\sigma_1 = 1.2 \cdot 10^{-3}$–$1.2\,S\cdot m^{-1}$, $\sigma_2 = 1 \cdot 10^{-8}$ $S\cdot m^{-1}$, $\sigma_3 = 0.5\,S\cdot m^{-1}$, $\varepsilon_1 = 78 \cdot \varepsilon_v$, $\varepsilon_2 = 10 \cdot \varepsilon_v$, $\varepsilon_3 = 60 \cdot \varepsilon_v$, $R = 2\,\mu m$, $d = 10\,nm$.

Determining the optimum suspending medium conductivity in terms of device sensitivity is an important question. For practical operation, it would be convenient to pass cells through the measurement system using standard culture medium or buffered saline. Nevertheless, as it represents an important adjustable parameter it is essential to determine whether the measurement sensitivity would benefit from resuspension of the sample in a medium of different conductivity. With this in mind the different approaches of AC electrokinetics and AC dielectric spectroscopy are compared below.

4.4.1. AC Electrokinetics

In AC electrokinetic techniques such as DEP and ROT, the movement of a cell under the influence of an applied electric field is used to obtain information about the complex electrical properties of the cell. The magnitude of the observed displacement and related forces are a function to the cell's dielectric properties. For DEP and ROT experiments, sucrose is generally added to the solution (to control the osmolarity of the solution) when working at decreased conductivities ($\sim 1.2 \cdot 10^{-2}$ S m^{-1}). From the observation of the Clausius–Mossotti factor frequency dependence for the single-shell cell model in **Fig. 7.9**, it is clear that separation and measurement techniques based on AC electrokinetics benefit from a low conductivity of the suspending medium:

- The characteristic times to polarize the membrane from the outside or the inside are well separated in the spectrum.

- The resulting forces on the particles are larger due to the Clausius–Mausotti factor but also because larger electric fields can be applied without generating heating and convective flows.

- Positive and negative DEP can both be induced for different excitation frequencies.

In AC electrokinetic techniques, measurement or separation are based on the cell motile response to the electric field and thus uncontrolled pressure or thermally induced flow (and for smaller particles Brownian motion) can be considered as noise. For cell sized particles the observable measurement noise in terms of Brownian motion can be neglected at room temperature and clearly does not depend on the liquid conductivity.

In ROT, other sources of measurement uncertainty exist, although not directly related to solution conductivity. Rotation rate can be influenced by adhesion between the cell and substrate or by dipole–dipole interactions of nearby cells. At low frequency, ROT measurements are also complicated by electroosmosis effects whereby the electric field rotates the double layer and hence the fluid around the particle.

4.4.2. AC Dielectric Spectroscopy

In AC dielectric spectroscopy, it is more difficult to answer the question of medium composition based only on the electrical properties of the sample because the detection circuit plays an important role in the determination of the sensitivity of the instrument.

Although a low conductivity medium increases the separation between the two Maxwell-Wagner dispersions in the spectrum as shown in **Fig. 7.10**, the resistance of the detection volume could well reach the MΩ range. This has number of important consequences:

- Higher Johnson noise is expected for low conductivity media.

- Measurement in the MHz range circuits for high impedance values are difficult to implement; the electronic and sensor are much more sensitive to parasitic capacitances.

- The fraction of conductive current in the system is significantly reduced whereas the dielectric current increases.

- To evaluate the total drop in permittivity in a low conductivity medium, data have to be acquired over a frequency band about twice as broad to find the low and high frequency values of the suspension permittivity and conductivity.

- If the cell to be measured is only present in the measurement electric field for a few milliseconds, then measurements done at the lower end of the frequency spectrum will be not have time to reach equilibrium.

This comparison between the two approaches seems to indicate that for AC dielectric spectroscopy, using low conductivities is not as beneficial as it is in the case of AC electrokinetics.

5. Cell Manipulation in Streams

Isolation and positioning of cells in a microfluidic channel by means of micromanipulation is essential if one is to obtain consistent results from the device. To reproducibly measure the particle electrical properties the system must be able to transport the sample solution precisely to and through specific locations within the chip. By reducing the fluidic channel dimensions, sequential tracking of each particle passing through the system detection area is made possible.

The flow rate should be held constant over the duration of the experiment. Accurate speed control and the option to stop or reverse the flow are desirable functionalities and are relatively easy to implement using external pressure or flow controllers. By varying the cross-section of the flow channel consecutive system, components (e.g., DEP focusing, impedance detection, sorting) can be integrated along the pathway that the cells take through the device, each requiring locally different flow and particle speeds.

A number of liquid pumping mechanisms applicable to microsystems are described in the literature, which can be separated into bulk and surface effects. Pressure-driven flow (PDF) is the simplest implementation for bulk techniques. The driving pressure can be generated outside of the chip with pumps or pressure regulators connected by short lengths of small bore tubing to minimize the compliance of the fluidics of the system as well as maintain sufficient flow in the sample supply tube to minimize sedimentation effects. A simple and precise method consist of priming the system and adjusting the height of a water column either up-stream or down-stream of the chip. Other techniques have been developed to exert pressure directly on the liquid inside the chip using piezoelectric pumping (35), centrifugal force (36), or thermal expansion (37).

Microfabricated chips allow the use of negative dielectrophoresis (nDEP) as a method to control the trajectories of particles and cells within the fluidic microchannel. Pairs of planar, overlapped top and bottom electrode strips are used to produce high electric field gradients within the liquid. The resulting force on the particle is proportional to the gradient of the electric field intensity squared and to the particle polarizability. In nDEP particles tend to move toward regions of lower electric field intensity. The electrodes are designed with a defined angle to the flow direction (**Fig. 7.11**). When entering the locally generated electric field, the particles moving under the influence of the PDF are subject to the nDEP

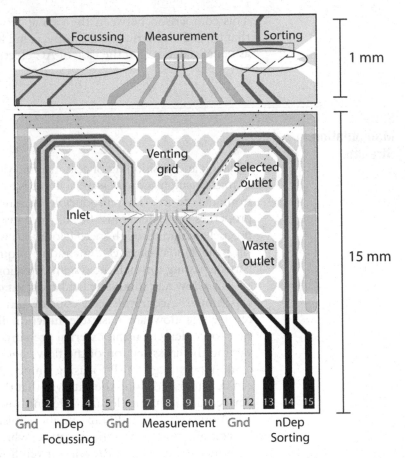

Fig. 7.11. Schematic of the chip bottom half structure showing the different electrodes functions and the electric and fluidic interface. A number of electrodes are unexploited or redundant in the impedance spectroscopy design, but can be used in other applications.

force acting perpendicular to the electrode strips. The Stokes force, from the fluid, can be separated into two components. The first, parallel to the electrodes, tends to move the particles along the electrodes edge. The second, perpendicular to the strip electrode, tends to push the particle into the electric field gradient and oppose the DEP force until the particle arrives at an equilibrium position a certain distance from the electrode edge. If the Stokes force is too large and force equilibrium is not reached, the particle will enter the electric field deeper and potentially cross the nDEP barrier.

A discussion of the nDEP and Stokes force on a particle as a function of the barrier angle θ to the fluid flow is given by Dürr et al.(38) and Schnelle et al. (39), using the expressions

$$F_{\text{DEP}} = \frac{27}{32} \pi^2 \varepsilon_1 \text{Re}[\tilde{K}] R^3 \frac{U_{\text{rms}}^2}{h^3}, \qquad [13]$$

$$F_{\text{Stokes}} = 6\pi \eta R v. \qquad [14]$$

The particles are deflected by the barrier for $F_{\text{DEP}} > F_{\text{Stokes}} \sin\theta$, which gives a maximal speed v_{max}

$$v_{max} = \frac{9\pi\varepsilon_1 \operatorname{Re}[\tilde{K}]}{64\eta} \frac{R^2 U_{rms}^2}{h^3 \sin\theta},$$ [15]

where R is the particle radius, U_{rms} is the root mean square of the electric voltage applied between the electrodes, and h is the channel height.

In current chip designs, the fluidic channel is wider in the nDEP regions (100–200 μm) than in the detection part of the system in order to reduce the flow speed and applied nDEP electric field. The nDEP electrode are 8 μm wide, which is a compromise between limitations of fabrication technology, the need for a high electric field gradient, and minimization of Joule heating in the channel.

A vertical component of the DEP force is also observed. Its effect is particularly visible on particles whose trajectories are initially close to the bottom surface of the channel. As the particles get close to the bottom electrode edge, they are swiftly pushed upward in the flow and their speed increases as they enter faster flow lines. As they move to faster lines, the particles experience a larger Stokes force and enter the barrier field more deeply and thus end up more centered. The final particle equilibrium position and particularly its height in the channel is thus a function of the Stokes, sedimentation, and nDEP forces. The estimated time constant for this effect is observed for the acceleration of slow particles as they move from the channel walls to the central flow axis and is on the order the order of a 100 ms. It should be in principle similar to what is observed in DEP-FFF (40).

Figure 7.11 shows the layout of a typical microfluidic chip including a DEP focusing area followed by a pair of measurement electrodes placed between ground shielding electrodes in a narrower channel section. A set of electrodes is added at the exit of the measurement channel to allow sorting of the particles toward one fluidic outlet or the other based on their measured properties (41).

To accommodate different chip fluidic configurations, the fluidic access to the chip or "holes" are defined on a grid with the possibility for three inlets and three outlets. **Figure 7.12** shows dual inlet and triple inlet configurations illustrating the flexibility of the fluidic system.

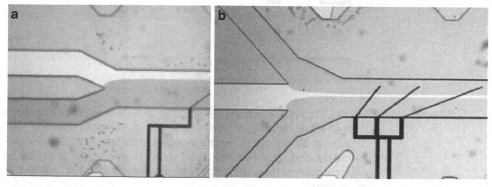

Fig. 7.12. Dual input (**a**) and sheath flow (**b**) options demonstrating the flexibility of the fluidic setup for pressure-driven flow control. (Images courtesy of N. Demierre and U. Seger.)

6. Chip Fabrication Process

Several techniques may be used to produce capillary channels in the micrometer size range, but in order to be able to accurately position electrodes on opposed channel walls the aligned sandwich technique was used.

Glass substrates were used (100 mm in diameter and 700 μm thick, Schott-Guinchard, Switzerland). The wafers are optically polished to $\lambda/10$ for $\varnothing$ 5 mm and the surface roughness is specified below 5 Å.

The first deposited layer consists of a thin metal film structure which is patterned by lift-off using either a reversible photoresist the AZ-5214 (Clariant) or the ma-N1440 (micro resist technology) which does not require an inversion bake (**Fig. 7.13a**). The sputtered electrode material consists of tantalum or titanium (20 nm), as an adhesion layer, and platinum (200 nm) as the active electrode material (**Fig. 7.13b**). Both layers were sputtered in vacuum using a Balzers 450. The deposition rate is set to ~0.3 nm s^{-1}. Lift-off is done in an ultrasonic acetone bath at 25°C, followed by an isopropanol rinse (**Fig. 7.13c**).

A thick photosensitive polyimide precursor (PI 2729, Dupont) is used to define the channels. It is spun over the patterned electrodes at a final speed of 4,000 rpm. A preliminary

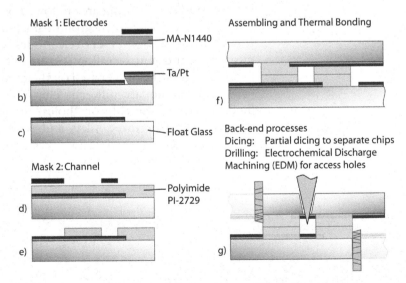

Fig. 7.13. Process flow for the microfluidic chip. (**a–c**) Patterning of the microelectrodes by lift-off. (**d–e**) Patterning of channel walls with photosensitive polyimide precursor. (**f**) Sealing of the channels by flip-chip alignment of the patterned wafer structures and thermal bonding. (**g**) Double side partial dicing to separate the chips and provide electrode contact access. Electrochemical discharge machining is used to open the access holes.

adhesion promoter dipping (VM-651, Dupont) is used to prevent edge detachment of the polyimide layer at development. A soft bake step of 5 min at 60°C followed by a 5 min at 100°C is performed prior to the layer exposure. Photolithography is done in hard contact mode and the exposure parameters are 40 s at 10 mW cm^{-2} (MA-6, Karl-Suss, **Fig. 7.13d**). The polyimide development process (**Fig. 7.13e**) is repeated three times in a cycle of three consecutive baths of 100% developer (DE6180, Dupont), 50–50% developer and rinse mixture and 100% rinse (RI9180, Dupont), 30 s in each bath and maintaining constant agitation. A final rinse is performed in a fresh rinse solution and development is checked optically. As two identical wafers will be put face-to-face to build a complete chip, the polyimide layer on each chip represents half of the final channel height.

The alignment of the two facing wafers is performed with a bond aligner (BA-6, Karl-Suss), with an estimated alignment precision of 2–3 μm. The thermal bonding (SB-6, Karl-Suss, **Fig. 7.13f**) of the facing polyimide layers is done under a tool pressure of 5 bars and by ramping the temperature to 200°C for a bake of 30 min in air and followed by 300°C in N$_2$ for 1h (**Fig. 7.14**). The first bake is performed in order to degrade the photo-package with oxygen, as it is known to migrate to the polyimide surface and prevent adhesion between the two layers. At higher temperature, an inert atmosphere is maintained to prevent oxidation of the polyimide, which is otherwise observed as darkening of the polymer layer after bonding.

Another aspect of the polyimide–polyimide bonding process is the significant degassing and consequent volume reduction that occurs during curing. The thickness of the layer after the thermal

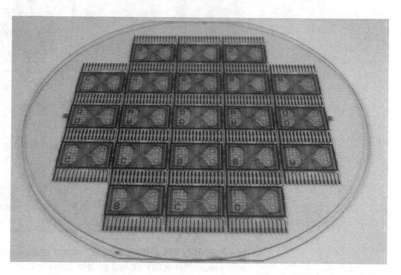

Fig. 7.14. Two bonded wafers containing 21 chips. Dicing and drilling are then performed to obtain individual functional devices.

bonding procedure is only 50% of the originally spun precursor. Degassing is responsible for the formation of bubbles at the interface between the polyimide layers. To allow these byproducts to vent, a venting grid is used around the fluidic channels. The design of this grid is such that measurement and DEP electrodes are covered by polyimide everywhere except in the channel and electrical contact areas. The final operations consist of the back-end processes performed in a grey room and include partial chip dicing in order to access the electrode contacts on each side of the chip and electrochemical discharge machining to drill the fluidic access holes (**Fig. 7.13 g**) (42).

The electrochemical access hole drilling process is shown in **Fig. 7.15**. With the wafer scale polyimide bonding, this operation represents one of the most critical step in the chip manufacturing

a

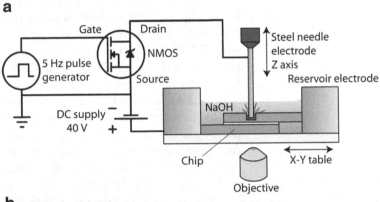

b

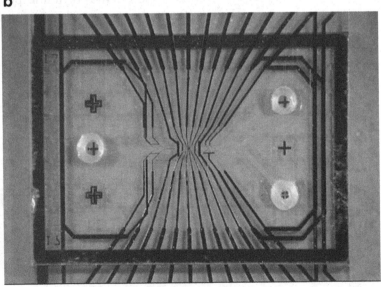

Fig. 7.15. Principle of glass electrochemical etching (**a**) and view of a microfluidic chip with electrochemically etched access holes (**b**). The chip is placed in a 30% NaOH solution. Electrical discharges are produced at a tip of a steel electrode which etches the glass at a rate of $\sim$20 $\mu m \cdot s^{-1}$.

as glass debris and other particles can enter and obstruct the microfluidic channels. Ultrasonic and PDF cleaning using different liquid viscosities can be used in cases where debris has entered the channels and generally permits 100% recovery of the chips.

Alternative techniques would be required in order to industrialize this process, allowing mass production of such devices. A possible approach would be to define the channel walls on a single wafer and to drill holes on the second prior to bonding.

6.1. Experimental Setup

A fluidic block composed of two mechanical parts is used to hold the chip in the center of the amplification electronics circuit board **Fig. 7.16**. The bottom part of the block is screwed to the printed circuit board and allows for precise placement of the chip. Two interface connectors each with 15 gold plated spring contacts (Samtec, 1 mm pitch) are used to contact the top and bottom of the platinum electrodes. Compared to other techniques, such as wire bonding or soldering, replacing the chip in the setup is very fast, requires no tools, and gives excellent reproducibility.

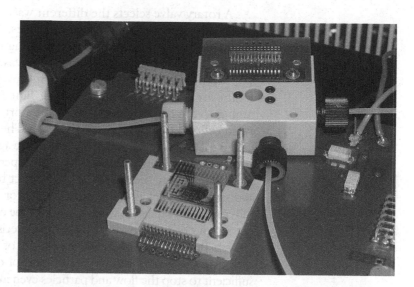

Fig. 7.16. The chip is placed in the PEEK electro-fluidic interconnection block. The top and bottom electrical connections are provided by two rows of 15 spring-loaded contacts. Optical access and transmissive illumination are provided through holes in the printed circuit board and block.

The top part is made of PEEK, which is resistant to solvents and bleach (**Fig. 7.17**). It provides L-shaped conduits to the chip fluidic apertures and the top side electrical connections to the chip. The fluidic contact is made watertight using miniature o-rings. The tubing connection to the fluidic block uses standard NPT 28 1/16 connectors. Additionally, an aperture is made in the center of the top block for non-fluorescent illumination of the chip.

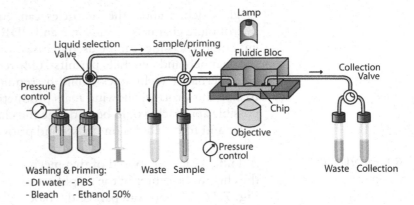

Fig. 7.17. Fluidic setup. The different washing liquids are either volume or pressure driven. The desired liquid or sample can be directed to the fluidic block through rotary valves. Miniature o-rings provide quick and water-tight fluidic connection to the chip. Waste vials are available before or after the chip for tube purging and fast sample injection.

A rotary valve selects the different washing or priming liquids, which are contained in pressurized bottles. Additionally, a motorized syringe pump is used for fast purging of the whole line with high pressure and proved very useful in case of channel clogging. The tubing inner diameter in the sample path is 120 μm, while it is 1 mm for the purging and washing liquids. The sample holder tubing to the fluidic block is made as short as possible in order to avoid cell sedimentation in the tubing. The holder supports standard 5 ml falcon tubes and provides sample agitation functionality to avoid sedimentation as well as optional temperature control. A valve is placed just before the fluidic block to select between cleaning liquid and sample. An inlet-side waste provides for a way to quickly purge the entry lines without having to remove the chip. Pressure control of the sample tube is provided by a high-precision pressure regulator (Marsh-Bellofram), with full scale pressure of 10 psi (0.7 bar), for ten turns. The regulator allows precise control of the fluid flow and is sufficient to stop the flow and particles even in the smallest part of the microchannel. The time constant of the system fluidics is estimated to be below 0.2 s. The height and level of the liquid in the collection vials is set so as to produce a small back-flow when atmospheric pressure is applied on the sample.

7. Measurements and Results

The primary aim of this section is to demonstrate that such a system can discriminate particles according to their dielectric and optical properties. We illustrate the efficacy of the system with the example of algae species discrimination. However, to test the system sensitivity, a number of preliminary experiments based on simple non-biological models are regularly performed. These steps are useful in defining or verifying a number of measurement parameters such as particles' flow speed, filters, or gain settings, which influence the measurements.

The calibration test is performed on the impedance spectroscopy instrument using beads of three different sizes, with diameters of 4.0 μm, and 6.0 μm (Duke scientific), and 5.14 μm (Molecular probes), which are mixed in the same sample tube. In **Fig. 7.18**, they are designated as 4, 5.14, and 6 μm beads. The coefficient of variation (CV) of the particles measured with a commercial FACS,

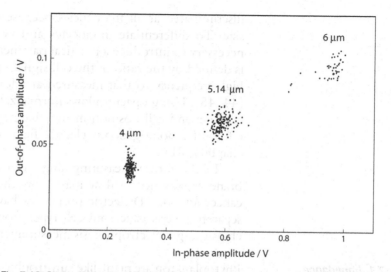

Fig. 7.18. Size-calibrated bead data as measured in PBS at 3 MHz using the integrated impedance sensor. Simple analysis of the signal shape of the dots outside the three main clouds showed they were due to doublets and were discarded for CV calculations.

the microfabricated impedance chip and those given by the bead manufacturers are summarized in **Table 7.1**.

Sub-micron size differentiation using calibrated polystyrene beads is thus easily achieved with such a system.

A number of measurements have previously demonstrated the application of such a system in the field of haematology and immunology. Published work also shows differentiation between red blood cells and red blood cell ghosts fixed with glutaraldehyde at various concentrations (43).

More recently, preliminary data using the impedance spectroscopy flow cytometry system shows the possibility of discrimination between neutrophils (granular, 12–15 μm diameter, multi-lobed nucleus), lymphocytes (5–10 μm, large nucleus), and monocytes (12–18 μm, U-shaped nucleus). Lymphocytes can be

Table 7.1
Summary of the size-calibrated bead measurements and manufacturer CV data. The precision FACS operating speed (100 events/s) permits accurate size resolution but is much slower than the normal sorting speed (~1,000 events/s)

Mean bead diameter (μm)	CV (manufacturer)	CV (FACS, 100/s)	Measured mean (3 MHz)	Measured CV (3 MHz)
4.0 (Duke)	1.15%	3.12%	0.2805 V	4.1%
5.14 (Duke, fluo.)	5.0%	3.14%	0.5885 V	4.2%
6.0 (M.P.)	0.42%	2.75%	0.9730 V	3.0%

discriminated at all frequencies because of their relatively small size. To differentiate monocytes and neutrophils, it was found necessary to introduce a practical parameter called opacity, which is defined by the ratio of the cell signal magnitude measured at a high frequency to that measured at a lower reference frequency (44, 45). Using opacity allows normalization of the data for both cell size and cell position in the channel (important due to the effect of inhomogeneous electric fields described earlier in the chapter) (31).

In the future measuring single-cell conductivity and membrane capacitance in flow may allow the rapid identification of cancerous cells. Dielectric properties have already been used to separate several cancerous cell types from normal blood cells in chips using dielectrophoresis measurement (46).

7.1. Impedance Spectroscopy and Optical Measurement of Algae

Phytoplankton are plant-like autotrophic microorganisms that live in the oceans and are at the base of the marine food chain. Analysis of phytoplankton density and species can be used to evaluate the effects of contaminants in marine waters. These contaminants can lead either to poisoning of the phytoplankton population or conversely to the formation of algal blooms that can create large amounts of toxins and rapidly sequester dissolved oxygen, causing problems to other marine life. Toxins from algal blooms are responsible for the death of the marine fauna and can cause disease in humans (47).

Phytoplankton are taxonomically and functionally diverse. They exist in a wide range of sizes (1–200 μm) and have biogeochemical characteristics that depend on their taxonomy. Measurement and analysis of phytoplankton taxa in the oceans is important for a number of reasons, including pollution monitoring (47), climate change (48), and climate modeling (49). A recent review by Legendre et al.(50) describes the analysis of phytoplankton in biological oceanography. These authors describe the contribution made by flow cytometric analysis to the understanding of the four major environmental crises; contamination of near-shore water, collapse of marine resources, loss of biodiversity, and global climate change.

The phytoplanktonic species presented here are *Isochrysis galbana*, *Synechococcus* sp., and *Rhodosorus m.* Cultures of these three algae were purchased from the Plymouth Algal Culture Collection (Marine Biological Association, UK). Batch cultures were grown in 250 ml conical flasks in a standard growth medium F/2 (51), illuminated on a 12/12 hour light/dark cycle at a light intensity of 30–50 μmol photons $s^{-1}m^{-2}$ at a temperature of 19°C. Sub-culturing was carried out every 2 weeks and the cultures were grown for 2 weeks before analysis. The cell diameters were measured using bright field microscopy and found to be in the range 3–4.5 μm for *Isochrysis galbana*, 4.5–7.0 μm for *Rhodosorus m.*, and 0.5–2.0 μm for *Synechococcus* sp. Photographs of each organism, taken in bright field and fluorescence, are shown in **Fig. 7.19.**

A mixture containing all three organisms was prepared by suspending the organisms in suspending medium F/2 at a concentration of 3 × 10^5 cells ml^{-1}. The mixture was then filtered with an 11-μm filter (NY11 Millipore) and a sample of approximately 10 μl was flowed through the microchip. The fluorescence from the particles was detected simultaneously through two emission filters to measure orange and red fluorescence **Fig. 7.20.** At the same time, the electrical impedance was measured at two frequencies (327 KHz and 6.030 MHz). The data acquisition software automatically registers an event when the fluorescent

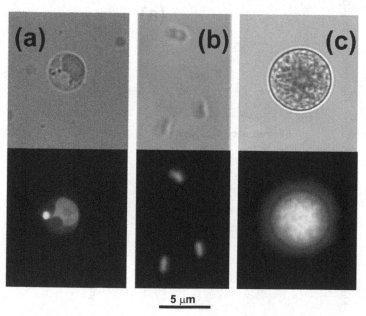

Fig. 7.19. Photographs of *I. Galbana* (**a**), *Synechococcus m.* (**b**), *Rhodosorus m.* (**c**). The upper photographs are taken in bright field while the bottom pictures were taken by exciting the cells with blue light (488 nm).

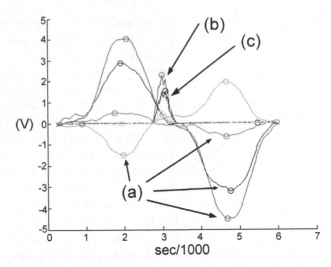

Fig. 7.20. A typical signal form a single algae cell. (**a**) Impedance signals (real and imaginary components) measured when the particle flowed across the two pairs of electrodes. The first set of peaks is the signal measured by the first pair of electrodes and the second peaks by the second pair; (**b**) fluorescence signals measured at center wavelengths of 675 nm; (**c**) fluorescence signals measured at center wavelengths of 675 nm and 585 nm.

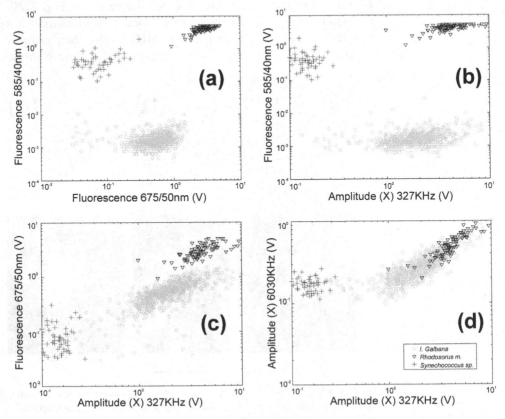

Fig. 7.21. Scatter plots in logarithmic scale for a typical experiment consisting of a sample of approximately 2,500 cells. (**a**) Magnitude of the orange fluorescence (585/40 nm) plotted against the red fluorescence (675/50 nm). The data show

light intensity in any optical channel exceeds a pre-set threshold level. For these experiments, the red fluorescent light signal was used as the trigger because it is emitted by all three cells.

The scatter plots in **Fig. 7.21(a–d)** show results from a typical experiment performed on a sample of approximately 2,500 cells. The three algae populations were mixed in a ratio of 50% *Synechococcus* sp., 15% *Rhodosorus m.*, and 35% *I. galbana* and re-suspended in F/2 medium, with electrical conductivity of 6.2 S m^{-1} (equivalent to sea water). The separation of the different cells population using the microfabricated device was found to be at least comparable but often better than that achieved with bulky and expensive commercially available flow cytometry systems such as the FACSAria.

8. Conclusion

A number of advantages and disadvantages arise from the miniaturization of the Coulter apparatus, requiring that a number of palliative steps (or slightly different approaches) be used to optimize the performance of the microfabricated impedance measurement device. Aspects of the electrical and physical theory behind the design and realization of the device have been explained, in some detail, to facilitate an understanding of the system operation. A system for simultaneous detection of optical and electrical properties of cells has been described and methods for its validation are discussed, as well as its application to the measurement of fluorescent algae.

For point-of-care applications, the possibility to measure single-cell dielectric properties with no other sample preparation other than a simple dilution presents definite advantages. For research applications involving cell-based assays, an integrated label-free diagnostic system presents interesting possibilities: limited number of preparation steps, faster results, and smaller sample volumes. In addition, as the cells are not modified by the detection

Fig. 7.21. (continued) that the three different populations can be easily resolved on the basis of their fluorescence emission spectrum. *I. galbana* emits in the red but not emit in the orange region, while *Rhodosorus m.* emits strongly in both the orange and the red spectrum. *Synechococcus sp.* can be identified because it emits both in the red and orange spectrum at very low intensity; (**b**) Orange fluorescence (585/40 nm) plotted against the real part of the impedance, (X-component), measured at a frequency of 327 kHz; (**c**) Red fluorescence (675/40 nm) plotted against the real part of the impedance (X-component), measured at a frequency of 327 kHz; either (**b**) and (**c**) show how the three populations of algae can be discriminated according to the level of fluorescence light emitted. (**d**) The real part of the impedance signal (measured at a 327 kHz) plotted against the real part of the impedance measured at 6.030 MHz on a log scale. At these frequencies, the electrical impedance measures the size of the particles; *Rhodosorus m.* is the larger organism and produces the biggest impedance signal. The electrical signal scales with size, i.e., *I. galbana* then *Synechococcus* sp.

process, further studies can be carried on the sample. The fact that the measurement is performed in physiological or culture medium suggests that cells could be cultivated, stored, and measured within similar types of chips.

The combination of the single-cell optical and impedance spectroscopy instrument with new integrated tools for cell manipulation, modification, and cultivation could lead to a variety of new flexible platforms for specific and general cell study.

Acknowledgment

The authors would like to acknowledge the contribution of several people without whose help completion of this chapter would have been impossible. Matt Mowlem from the National Oceanographic Centre, Southampton, for his assistance with all aspects of the algae work; Nicolas Green and Sun Tao for all the discussions on dielectric theory; Pontus Linderholm, Urban Seger, and Thomas Braschler for useful discussions on electrode geometries, configurations, and electronics. Raphael Tornay for his collaboration on the data acquisition software and the CMI team at EPFL for their support in fabricating the chips. Also the authors would like to express their thanks to Börge Wessling of the IWE, RWTH Aachen University for discussions on nanoporous electrodes and performing the IrOx sputtering on the microelectrodes. Shady Gawad would like to thank Hywel Morgan for the invitation to spend a wonderful year in his lab and would also like to acknowledge the financial support of the FNRS "Bourse pour jeunes chercheurs."

References

1. Li, Q. J., Dinner, A. R., Qi, S. Y., Irvine, D. J., Huppa, J. B., Davis, M. M. and Chakraborty, A. K. (2004) CD4 enhances T cell sensitivity to antigen by coordinating Lck accumulation at the immunological synapse. *Nature Immunology* **5**, 791–799.

2. Wee, S. F., Schieven, G. L., Kirihara, J. M., Tsu, T. T., Ledbetter, J. A. and Aruffo, A. (1993) Tyrosine Phosphorylation of Cd6 by Stimulation of Cd3 – Augmentation by the Cd4 and Cd2 Coreceptors. *Journal of Experimental Medicine* **177**, 219–223.

3. Fu, A. Y., Chou, H. P., Spence, C., Arnold, F. H. and Quake, S. R. (2002) An integrated microfabricated cell sorter. *Analytical Chemistry* **74**, 2451–2457.

4. Kruger, J., Singh, K., O'Neill, A., Jackson, C., Morrison, A. and O'Brien, P. (2002) Development of a microfluidic device for fluorescence activated cell sorting. *Journal of Micromechanics and Microengineering* **12**, 486–494.

5. Schrum, D. P., Culbertson, C. T., Jacobson, S. C. and Ramsey, J. M. (1999) Microchip Flow Cytometry Using Electrokinetic Focusing. *Analytical Chemistry* **71**, 4173–4177.

6. Davey, H. M. and Kell, D. B. (1996) Flow cytometry and cell sorting of heterogeneous microbial populations: The importance of single-cell analyses. *Microbiological Reviews* **60**, 641–&.

7. Ayliffe, H. E., Frazier A. B. and Rabbitt, R. D. (1999) Electric Impedance spectroscopy

using microchannels with integrated metal electrodes. *IEEE Journal of Microelectromechanical Systems* **8**, 50–57.

8. Pohl, H. A. (1978) *Dielectrophoresis.* Cambridge University Press, Cambridge, UK.

9. Gimsa, J. (1999) Particle characterization by AC-electrokinetic phenomena: 1. A short introduction to dielectrophoresis (DP) and electrorotation (ER). *Colloids and Surfaces a-Physicochemical and Engineering Aspects* **149**, 451–459.

10. Wang, X. B., Huang, Y., Gascoyne, P. R. C. and Becker, F. F. (1997) Dielectrophoretic manipulation of particles. *IEEE Transactions on Industry Applications* **33**, 660–669.

11. Fuhr, G., Glaser, R. and Hagedorn, R. (1986) Rotation of dielectrics in a rotating electric high-frequency field. Model experiments and theoretical explanation of the rotation effect of living cells. *Biophysical Journal* **49**, 395–402.

12. Huang, Y., Wang, X. B., Gascoyne, P. R. C. and Becker, F. F. (1999) Membrane dielectric responses of human T-lymphocytes following mitogenic stimulation. *Biochimica Et Biophysica Acta-Biomembranes* **1417**, 51–62.

13. Yang, J., Huang, Y., Wang, X. J., Wang, X. B., Becker, F. F. and Gascoyne, P. R. C. (1999) Dielectric properties of human leukocyte subpopulations determined by electrorotation as a cell separation criterion. *Biophysical Journal* **76**, 3307–3314.

14. Gascoyne, P. R. C., Mahidol, C., Ruchirawat, M., Satayavivad, J., Watcharasit, P. and Becker, F. F. (2002) Microsample preparation by dielectrophoresis: Isolation of malaria. *Lab on a Chip* **2**, 70–75.

15. Jones, T. B. (1995) *Electromechanics of Particles.* Cambridge University Press, Cambridge, p. 285.

16. Morgan, H. and Green, N. G. (2003) *AC Electrokinetics: Colloids and Nanoparticles.* Research Studies Press Ltd, Baldock.

17. Asami, K., Takahashi, Y. and Takashima, S. (1989) Dielectric properties of mouse lymphocytes and erythrocytes. *Biochimica et Biophysica Acta (BBA) – Molecular Cell Research* **1010**, 49–55.

18. Asami, K., Yonezawa, T., Wakamatsu, H. and Koyanagi, N. (1996) Dielectric spectroscopy of biological cells. *Bioelectrochemistry and Bioenergetics* **40**, 141–145.

19. Gheorghiu, E. and Asami, K. (1998) Monitoring cell cycle by impedance spectroscopy: Experimental and theoretical aspects. *Bioelectrochemistry and Bioenergetics* **45**, 139–143.

20. Coulter, W. H. US Patent 2,656,508. Means for counting particles suspended in fluid, 1953.

21. Larsen, U. D., Blankenstein, G. and Ostergaard, S (1997) In *Microchip Coulter Particle Counter*, Proceedings of Transducers, Chicago, USA, pp. 1319–1322.

22. Koch, M., Evans, A. G. R. and Brunnschweiler, A. (1998) In *Design and Fabrication of a Micromachined Coulter Counter*, Proceedings of Micromechanics Europe, Ulvik, Norway, pp. 155–158.

23. Koch, M., Evans, A. G. R. and Brunnschweiler, A. (2000) *Microfluidic Technology and Applications.* Research Studies Press Ltd, Baldock.

24. Fuller, C. K., Hamilton, J., Ackler, H. and Gascoyne, P. R. C. (2000) In *Microfabricated Multi-Frequency Particle Impedance Characterization System*, Proceedings of Micro Total Analysis Systems, Enschede, Netherlands, pp. 265–268.

25. Larsen, U. D., Norring, H. and Telleman, P. (2000) In *Somatic Cell Counting with Silicon Apertures*, Proceedings of Micro Total Analysis Systems, Enschede, Netherlands pp. 103–106.

26. Gawad, S., Heuschkel, M., Leung-Ki, Y., Iuzzolino, R., Schild, L., Lerch, P. and Renaud, P.(2000) In *Fabrication of a Microfluidic Cell Ananlyzer in a Microchannel Using Impedance Spectroscopy*, Proceedings of the IEEE-EMBS Conference on Microtechnologies in Medicine & Biology, Lyon, France, pp. 297–301.

27. Onaral, B. and Schwan, H. P. (1982) Linear and non-linear properties of platinum-electrode polarization. 1. Frequency-dependence at very low-frequencies. *Medical & Biological Engineering & Computing* **20**, 299–306.

28. Franks, W., Schenker, W., Schmutz, P. and Hierlemann, A. (2005) Impedance characterization and modeling of electrodes for biomedical applications. *IEEE Transactions on Biomedical Engineering* **52**, 1295–1302.

29. Wessling, B., Mokwa, W. and Schnakenberg, U. (2006) RF-sputtering of iridium oxide to be used as stimulation material in functional medical implants. *Journal of Micromechanics and Microengineering* **16**, S142–S148.

30. Kachel, V. (1990) Electrical resistance pulse sizing: Coulter sizing. In *Flow Cytometry and*

Sorting (second edition), Melamed, M. R., (Ed.). Wiley-Liss, New York, pp. 45–80.

31. Gawad, S., Cheung, K., Seger, U., Bertsch, A. and Renaud, P. (2004) Dielectric spectroscopy in a micromachined flow cytometer: Theoretical and practical consideration. *Lab on a Chip* **4**, 241–251.

32. Braschler, T., Johann, R., Seger, U., Linderholm, P., Demierre, N. and Renaud, P. (2005) In *Single Cell Positioning, Entrapment and Electrical Characterisation*, Proceedings of MicroTAS, Boston, USA, pp. 403–405.

33. Holmes, D., Green, N. G. and Morgan, H (2003) Microdevices for Dielectrophoretic flow-through separation. *IEEE Engineering in Medicine and Biology Magazine* **22/6**, 85–90.

34. Pauly, H. and Schwan, H. P. (1959) Uber Die Impedanz Einer Suspension Von Kugelformigen Teilchen Mit Einer Schale – Ein Modell Fur Das Dielektrische Verhalten Von Zellsuspensionen Und Von Proteinlosungen. *Zeitschrift Fur Naturforschung Part B-Chemie Biochemie Biophysik Biologie Und Verwandten Gebiete* **14**, 125–131.

35. van Lintel, H., van de Pol, F. and Bouwstra, S. (1988) A piezoelectric micropump based on micromachining of silicon. *Sensors and Actuators A: Physical* **15**, 153–168.

36. Kellogg, G. J., Arnold, T. E., Carvalho, B. L. and Duffy, D. C. (2000) In *Centrifugal Microfluidic: Applications*, Proceedings of Micro Total Analysis Systems, Enschede, Netherlands, pub. Kluwer, pp. 239–242.

37. Jun, T. K. and Kim, C. J. (1998) Valveless pumping using traversing vapor bubbles in microchannels. *Journal of Applied Physics* **83**, 5658–5664.

38. Dürr, M., Schnelle, T., Müller, T. and Stelzle, M. (2001) In *Dielectrophoretic Separation and Accumulation of Bio Particles in Micro-Fabricated Continuous Flow Systems*, Proceedings of Micro Total Analysis, Monterey, USA, pub. Kluwer, pp. 539–540.

39. Schnelle, T., Muller, T., Gradl, G., Shirley, S. G. and Fuhr, G. (2000) Dielectrophoretic manipulation of suspended submicron particles. *Electrophoresis* **21**, 66–73.

40. Yang, J., Huang, Y., Wang, X. B., Becker, F. F. and Gascoyne, P. R. C. (2000) Differential analysis of human leukocytes by dielectrophoretic field-flow-fractionation. *Biophysical Journal* **78**, 2680–2689.

41. Gawad, S., Batard, P., Seger, U., Metz, S. and Renaud, P. (2002) In *Leukocytes Discrimination by Impedance Spectroscopy Flow Cytometry*, Proceedings of Micro Total Analysis Systems, Nara, Japan, pub. Kluwer, pp. 649–651.

42. Fascio, V., Langen, H. H., Bleuler, H. and Comninellis, C. (2003), Investigations of the spark assisted chemical engraving. *Electrochemistry Communications* **5**, 203–207.

43. Cheung, K., Gawad, S. and Renaud, P. (2005) Impedance spectroscopy flow cytometry: On-chip label-free cell differentiation. *Cytometry Part A* **65A**, 124–132.

44. Hoffman, R. A. and Britt, W. B. (1978), Flow-system measurement of cell impedance properties. *Journal of Histochemistry and Cytochemistry* **27**, 234–240.

45. Hoffman, R. A., Johnson, T. S. and Britt, W. B. (1981) Flow cytometric electronic direct-current volume and radiofrequency impedance measurements of single cells and particles. *Cytometry* **1**, 377–384.

46. Gascoyne, P. R. C., Wang, X. B., Huang, Y. and Becker, F. F. (1997) Dielectrophoretic separation of cancer cells from blood. *IEEE Transactions on Industry Applications* **33**, 670–678.

47. Graham, L. E. and Wilcox, L. W. (2000) *Algae*. Prentice Hall, Upper Saddle River.

48. Nehring, S. (1998) Establishment of thermophilic phytoplankton species in the North Sea: Biological indicators of climatic changes? *ICES Journal of Marine Science* **334**, 340–343.

49. Sarmiento, J. L. and Le Quéré, C. (1996) Oceanic carbon dioxide uptake in a model of century-scale global warming. *Science* **274**, 1346–1350.

50. Legendre, L., Courties, C. and Troussellier, M. (2001) Flow cytometry in oceanography 1989–1999: Environmental challenges and research trends. *Cytometry* **44**, 164–72.

51. Guillard, R. R. L. (1975) Culture of phytoplankton for feeding marine invertebrates. In *Culture of Marine Invertebrate Animals*, Smith, W. L. and Chanley, M. H., (Eds.). Plenum Press, New York, USA, pp. 26–60.

Chapter 8

Dielectrophoresis as a Cell Characterisation Tool

Kai F. Hoettges

Abstract

Dielectrophoresis (DEP) is a technique which offers label-free measurement of cell electrophysiology by monitoring its movement in non-uniform electric fields. In this chapter, the theory underlying DEP is explored, as are the implications of the development of equipment for taking such measurements. Practical considerations such as the selection of a suspending medium are also discussed.

Key words: AC electrokinetics, microelectrodes, dielectrics.

1. Dielectrophoresis

Dielectrophoresis (DEP) is the motion of dielectric particles in a non-homogeneous electric field, and was first observed by Pohl in 1951 (1). In contrast to electrophoresis, which affects only charged particles, DEP affects all particles that have different dielectric properties than the medium they are suspended in. A homogenous electric field induces a dipole in the particle and leads to coulomb forces acting on both sides of the particle. However, since the field is homogenous, the forces on both will cancel out so no net force is exerted on the particle. The inhomogeneous field gradient used in DEP polarises the particle to a different extent on opposing sites of the particle, so the strength of dipole varies across the particle and thereby exerts a force on the particle. Since the dipole is induced by an inhomogeneous electric field, the dipole will be stronger where the field gradient is strongest, independent of the field polarity, which has no influence on the direction of the force (**Fig. 8.1**). This allows the use of AC fields for DEP, offering the additional advantage that electrophoresis (based on net

M.P. Hughes, K.F. Hoettges (eds.), *Microengineering in Biotechnology*, Methods in Molecular Biology 583,
DOI 10.1007/978-1-60327-106-6_8, © Humana Press, a part of Springer Science+Business Media, LLC 2010

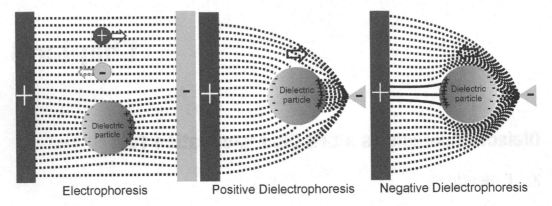

Fig. 8.1. In electrophoresis effects charged particles experience a force, dielectric particles polarise and distort the electric field but charges cancel out so there is no net force. In dielectrophoresis the electric field is highly inhomogeneous, so particles (and the surrounding medium) polarise to a different extent on opposite sites of the particle and the particle experiences a net force. If the particle is more polarisable then the medium it will move toward the field gradient (positive DEP). If it polarises less then the medium the particle is repelled from the field gradient (negative DEP).

charge) is eliminated. The use of AC fields also prohibits electro-chemical reactions such as electrolysis at the electrodes, and thereby reduces corrosion of the electrodes.

The strength and direction of the force depends on the polarisability of the particle in relation to the surrounding medium at a given frequency. If the particle polarises stronger than the medium, the particle will be attracted to the regions of higher field gradient (normally the electrode edge), while a particle that is less polarisable than the medium will be repelled from regions of high-field gradient.

Since the complex permittivity of the particle is frequency dependent, the same particle can experience positive or negative DEP at different frequencies. This can be used to characterise particles by recording the strength and direction of the DEP force against frequency and fitting a mathematical model of the particle against the recorded data. This will be covered in the following discussion in more detail.

There has been a significant body of work reported on the use of DEP for cell separation by using a frequency that attracts one type of cells and repels another in flow cells. This was demonstrated for separation of viable and non-viable yeast cells (2, 3) or healthy and leukemic blood cells (4). DEP was also used for cell manipulation by moving or levitating cells using negative DEP (5, 6).

However, this chapter will mainly focus on using DEP to characterise cells. To use DEP as a characterisation technique, the strength and direction of the DEP force is measured over a wide frequency range. A mathematical model of a particle can be fitted to the resulting DEP spectrum to extract the physical properties of the particle. Applied to cells, this allows the characterisation of the cells electrophysiology. DEP has several advantages over conventional techniques. First of all DEP is a label-free technique, while many other techniques use chemical labels such as fluorescent stains, to mark a parameter. Interactions between labels and cells are common so a large number of false positive or negative results are caused by these interactions. By avoiding the use of labels, these problems can be avoided. DEP is also a nondestructive technique, allowing use of the cells for further experiments after characterisation.

Examples of cells studied in the literature range from cancer cells (7–9), red blood cells (10), yeast cells (11), bacteria (12, 13), algae (14), virus particles (15), latex beads (16, 17) or carbon nanotubes (18). Further examples are reviewed in (19–21).

2. Measuring the Dielectrophoretic Response

To quantify the dielectrophoretic response, a number of techniques are available. Since it is impossible to directly measure the force on a particle, all these techniques rely on indirect measures such as the number of particles collected, the speed of motion or the levitation height. In general, they only give a proportional response of the strength and direction of the force that needs to be scaled to the model spectrum. This also means that the model does not need to calculate the absolute force on the particle, so most models consider only the Clausius–Mossotti factor of a particle. The most common techniques are discussed below.

2.1. Collection Rate Measurements

The most common as well as simplest technique is collection rate measurement and has been demonstrated for the characterisation of numerous cell lines (8, 9) as well as bacteria (12) and algae (14). Electrodes are submerged in a suspension with a known concentration of particles and the particles that collect at the electrode over a given period of time are counted either by simple counting or, in the case of fluorescent particles, by measuring the fluorescence at the electrode edge. To obtain a spectrum of the DEP force the experiment is repeated over a wide range of frequencies (usually on a logarithmic scale with three to seven data points per decade). Since particles are only counted at the electrode edge, collection rate measurements are only suitable to quantify positive DEP. Some electrode geometries have also defined minima where negative DEP could be quantified. However, since the geometries of the maxima and minima in the field gradient are different, the scale of the response is different for positive and negative DEP, so it is difficult to obtain continuous spectra which have a zero transition.

2.2. Crossover Measurements

The measurement of the crossover frequency between positive and negative DEP can also be used to characterise particles. It is mainly used for smaller particles, such as viruses (22–24), that are difficult to count with collection spectra. The most common way to collect crossover data is to trap particles using positive DEP and then sweeping the frequency until the particles are repelled from the electrodes. The change from positive to negative DEP indicates the crossover frequency. For a full characterisation, the experiments need to be repeated at different medium conductivities. A common problem with crossover measurements is particles adhering to the electrodes (and each other) are so strong that negative DEP cannot repel them. In this case only a small proportion is repelled by negative DEP, which makes the observation more difficult.

2.3. Particle Velocity Measurements

A further method is to measure the speed and direction of a particle in an electric field gradient (25, 26). The stronger the force, the faster the particle. The experiments are usually recorded and video-analysis algorithms are used to track particles from one frame to the next to monitor the speed of the particle. By measuring the speed of the particle in a known field gradient at different frequencies the DEP force can be quantified.

2.4. Measurement of the Levitation Height

A method for quantifying negative DEP is to monitor the levitation height of a particle over the electrode array (27, 28). Once the particle is in stable levitation, the buoyancy force of the particle and the DEP force should cancel out. So a particle expressing strong negative DEP will levitate higher over the array than a particle experiencing a weaker force. The technique is good for rapid assessment of single particles and is mainly used for larger particles such as cells. The main problem limiting this technique is the accuracy of the height measurement.

2.5. Impedance Sensing

Particles collecting at the electrode edge have an influence on the impedance of the electrodes. This change in impedance can be monitored to quantify DEP. Conductive particles build pearl chains across the inter-electrode gap. This reduces the impedance between the two electrodes. In many cases, the impedance of the particles itself is only responsible for a small part of the impedance change since the higher ion concentration in the electrical double layer around the particles contributes largely to the reduction in impedance. Impedance sensing is mostly used for very conductive particles such as carbon nanotubes (18, 29) but was also demonstrated on bacteria (30) and latex beads (31).

3. Modelling and Interpreting DEP Spectra

To interpret data gained from experiments, cell response at deferment frequencies can be matched to mathematical models. By matching a computer model to experimental data the properties of the particles can be determined.

3.1. Theory of Particle Modelling

The dielectrophoretic force (F_{DEP}) for a homogenous particle suspended in a local field gradient is given by

$$F_{DEP} = 2\pi r^2 \varepsilon_m \text{Re}[K(\omega)] \Delta E^2, \qquad [1]$$

where r is the radius of the particle, ε_m is the permittivity of the suspending medium and ΔE is the gradient of the rms electric field. $\text{Re}[K(\omega)]$ is the real part of the Clausius–Mossotti factor, which is given by

$$K(\omega) = \frac{\varepsilon_p^* - \varepsilon_m^*}{\varepsilon_p^* + 2\varepsilon_m^*}, \qquad [2]$$

where ε_m^* and ε_p^* are the complex permittivity of the medium and particle respectively. These are given by

$$\varepsilon^* = \varepsilon - \frac{j\sigma}{\omega}, \qquad [3]$$

where ε *is* the permittivity, σ the conductivity and ω *is* the angular frequency of the electric field. For practical particle characterisation only the real part of the Clausius–Mossotti factor $\text{Re}[K(\omega)]$ against frequency needs to be considered since it gives proportional strength and direction of F_{DEP}. The most common techniques to quantify F_{DEP} give only a proportional response that depends on many factors such as particle concentration, electrode geometry, voltage, etc. Therefore the results can be scaled to the model by multiplying it with a correction factor without loosing information about the particles' properties.

The Clausius–Mossotti factor can be calculated not only for homogeneous particles but also for more complex particles such as cells, by modelling them as concentric spheres (e.g. the cytoplasm surrounded by a membrane). The most common model was developed by Irimajiri (32) and considers each sphere as a homogenous particle suspended in a medium that has the properties of the next bigger sphere. Starting from the innermost sphere, we can calculate the dispersion at the interface between the inner and the outer of the two spheres. This combined particle can then be modelled as a homogeneous sphere inside the next larger one. In this way the properties of all individual spheres can be combined to give the total response of the entire particle. To model it mathematically we consider a particle of N concentric spheres forming a multi-layered particle. To each layer i we assign an outer radius a_i, where a_1 is the radius of the core and a_{N+1} is the radius of the outermost sphere, and thereby the radius of the entire particle. Each of these layers has its own complex permittivity given by

$$\varepsilon_i^* = \varepsilon_i - \frac{j\sigma_i}{\omega},$$ [4]

where i has values from 1 to $N+1$. To determine the effective properties for the whole particle, we first determine the properties of the core surrounded by the next layer in order to determine the effective properties of a homogeneous particle consisting of the core suspended in one sphere. In practice this could be considered as the cytoplasm of a cell surrounded by a membrane. This particle has a radius of a_2 and the effective permittivity is given by

$$\varepsilon_{1\text{eff}}^* = \varepsilon_2^* \frac{\left(\frac{a_2}{a_1}\right)^3 + 2\frac{\varepsilon_1^* - \varepsilon_2^*}{\varepsilon_1^* + 2\varepsilon_2^*}}{\left(\frac{a_2}{a_1}\right)^3 - \frac{\varepsilon_1^* - \varepsilon_2^*}{\varepsilon_1^* + 2\varepsilon_2^*}}.$$ [5]

We now repeat the procedure for an additional layer to model a particle consisting of three concentric spheres, by using the values calculated for the inner two spheres as a homogenous inner particle suspended in an additional sphere. This could be considered as the cytoplasm of a bacterium surrounded by a membrane and an outer cell wall:

$$\varepsilon_{2\text{eff}}^* = \varepsilon_3^* \frac{\left(\frac{a_3}{a_2}\right)^3 + 2\frac{\varepsilon_{1\text{eff}}^* - \varepsilon_3^*}{\varepsilon_{1\text{eff}}^* + 2\varepsilon_3^*}}{\left(\frac{a_3}{a_2}\right)^3 - \frac{\varepsilon_{1\text{eff}}^* - \varepsilon_3^*}{\varepsilon_{1\text{eff}}^* + 2\varepsilon_3^*}}.$$ [6]

By repeating the procedure a further $N-2$ times, we can calculate the effective permittivity $\varepsilon_{P\text{eff}}^*$ of a particle consisting of N concentric spheres:

$$\varepsilon_{P\text{eff}}^* = \varepsilon_{N\text{eff}}^* \frac{\left(\frac{a_{N+1}}{a_N}\right)^3 + 2\frac{\varepsilon_{N-1\text{eff}}^* - \varepsilon_{N+1}^*}{\varepsilon_{N-1\text{eff}}^* + 2\varepsilon_{N+1}^*}}{\left(\frac{a_{N+1}}{a_N}\right)^3 - \frac{\varepsilon_{N-1\text{eff}}^* - \varepsilon_{N+1}^*}{\varepsilon_{N-1\text{eff}}^* + 2\varepsilon_{N+1}^*}}.$$ [7]

This equation provides a value for the combined complex permittivity of the particle at a given angular frequency ω. Combined with the complex permittivity of the medium it can be used to calculate the Clausius–Mossotti factor over the frequency range that was used for experimental measurements. Other models for non-spherical geometry such as ellipsoids or rods are reported in the literature and can be used to model particles such as red blood cells (33), elongated bacteria (34) or carbon nanotubes (18).

Various algorithms and programs for computer models are published as well (35, 36). The model can be curve fitted to the data by adjusting the conductivity, permittivity and layer thickness for each layer. This allows extracting the electrophysiology of the particle. In the case of cells, the properties of the cytoplasm and membrane can give valuable information. Factors such as ion flux or even water efflux change the conductivity of the cytoplasm, while damage to the membrane or membrane folding change membrane parameters. This allows label-free investigation of ion channel activity, cell viability or interactions with drugs (**Fig. 8.2**). The quality of the model generally improves the more variables are known. Therefore it is advisable to measure the size of the cell with a microscope. Factors such as the membrane thickness stay reasonably constant (5–10 nm for a lipid bi-layer membrane) while factors such as cell wall thickness of bacteria can often be found in the literature. In case only a part of the cells is affected (e.g. in IC_{50} measurements), algorithms that evaluate overlapping spectra of multiple populations allow identifying the proportion of cells in each population (37).

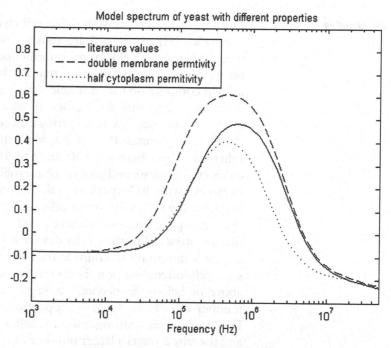

Fig. 8.2. Model spectrum of yeast. The solid line is modelled using a model consisting of three concentric spheres using published literature values (49). For the *dashed line* the membrane permittivity was doubled, while for the doted line the cytoplasm permittivity was halved. This illustrated that the cytoplasm conductivity affects mainly the high frequency end of the plot while membrane properties affect at which frequency the peak rises.

4. Electrode Geometries

To perform DEP experiments, electrodes that generate a high-field gradient are needed. While early electrodes applied high voltages using wires as electrodes, more recent work focussed on electrodes with inter-electrode gaps in the micrometer range. As a "rule of thumb" an inter-electrode gap of 30 times the particle diameter is considered a good starting point. For most cells electrodes are therefore spaced between 10 μm and 100 μm, but smaller gaps are used for particles such as viruses or nanoparticles. Inter-electrode gaps in this rage can be normally powered with standard bench-top signal generators that supply between 5 V and 20 V (peak to peak), while a frequency range between 500 Hz and 50 MHz is investigated. (Lower frequencies cannot reliably suppress electrochemical reactions at the electrodes while most "interesting" features for cell characterisation appear below 50 MHz). In general data are recorded on a logarithmic frequency scale with three to seven measurements per decade.

There are numerous electrode geometries described in the literature, the most common ones are summarised below.

4.1. Needles

The simplest electrode system for cell characterisation are needle electrodes (**Fig. 8.3**) and are mostly used to record collection spectra; however, velocity measurements are possible too. They offer a simple low cost solution that can be rapidly fabricated and used in conjunction with a simple signal generator, but can nevertheless be a very valuable tool for cell characterisation. They consist of two metal needles (e.g. syringe needles) fixed in a sample chamber (e.g. small Petri dish), with the tips pointing at each other and a gap between 100 μm and 400 μm. The needles are connected to phase and ground of a signal generator powering the electrodes with 20 V (peak to peak). For measurements the chamber is filled with a solution of cells (ca 3×10^5 cells/mL) and the signal is applied for a fixed time (e.g. 60 s). During this time all cells that are attracted to the electrodes are counted. (For higher accuracy, it is important to count as they move towards the electrode, since cells often disappear from view as they may attach out of view above or below the needle, leading to a lower cell count). The number of cells attracted is proportional to the strength of the DEP force, since a stronger force reaches further into the medium and thereby attracts a larger number of cells. After each frequency the signal generator is switched off and the cells are re-suspended by agitating the medium with a micropipette.

The experiments are repeated over a wide frequency range and the number of cells is plotted against frequency. The main drawback of this technique is that only positive DEP can be detected

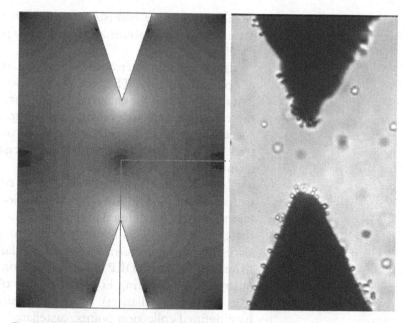

Fig. 8.3. Finite element model of the electric field gradient (*left*) around needle electrodes. High-field gradients (light) are around the tips of the needle but there is no defined minimum. Right image: Yeast cells collecting on needled electrodes (positive DEP).

using needles since negative DEP repels the cells into the medium, making it impossible to quantify the effect. Also data obtained with needle electrodes tend to be noisy since only a relatively small number of cells is counted during each experiment (up to 100 cells). Therefore, experiments normally need to be repeated several times and averaged to get a clear spectrum. Data can vary from device to device; since it is difficult to construct needles with a constant inter- electrode gap and geometry of the tip. Needles can also be used for cell velocity measurements, but this technique is less common for this geometry.

4.2. Micro-electrodes

A huge body of work was performed using micro-fabricated electrodes. There are a number of geometries that will be described here. In general they are all fabricated using micro fabrication techniques adapted from semiconductor manufacturing, and consist of thin metal films (mostly gold, chromium or titanium) on a flat insulating substrate (glass oxidised silicon or plastic). The metal film is patterned using photolithography and etched to form microelectrodes of various shapes.

In general microelectrodes generate a very high-field gradient since they consist of thin metal films (100–500 nm), so even broad electrode tracks generate high gradients at their edge that could not be achieved with thicker wire or needle electrodes.

The simplest geometry is again needle electrodes, whereby two triangular electrodes point at each other. The advantage over "handmade" needle electrodes is that microelectrodes can be fabricated more consistently and with smaller inter-electrode gaps. Also rows of points are common to perform experiments in parallel.

Interdigitated electrodes consist of parallel track of electrodes of opposite polarity. They are mainly used for separation or cell levitation and not for characterisation since it is difficult to define the electrode area to evaluate. However, they can be used for collection spectra (the cells are counted over a defined length of track) or velocity measurements. Again negative DEP is difficult to quantify.

Interdigitated castellated electrodes consist of parallel tracks with castellation (**Fig. 8.4**) (38). These electrodes have the advantage that there is a defined minimum of the field gradient between the castellations. This allows to verify negative DEP; however, since the minimum form a large triangular area it is still difficult to quantify negative DEP. Positive DEP, on the other hand, can be quantified by counting how many cells collect on a defined area (e.g. on one castellation). Since the positive and negative DEP have defined collection points, castellated electrodes can be used for crossover measurements.

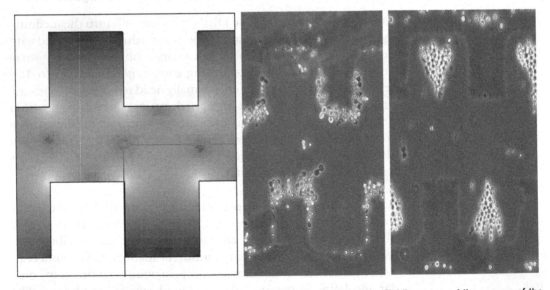

Fig. 8.4. Castellated electrodes. Finite element model in the left: The field maxima (light) are around the corners of the castellation so positive DEP will attract the cells to this while the field minimum forms a triangular region between two castellation where cells get pushed by negative DEP. The centre picture shows positive DEP of yeast while the right picture shows negative DEP of yeast.

Polynomial electrodes consist of four curved electrodes arranged in a square (**Fig. 8.5**). There are a number of geometries described in the literature (39–41), the most common being circular, parabolic or those based on the isomotive design described by Pohl (42). If a two-phase signal generator is used,

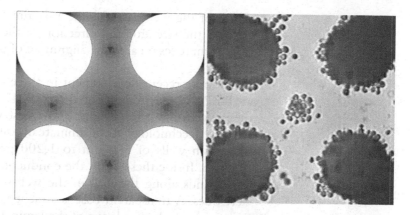

Fig. 8.5. Polynomial electrodes have a maximum (*light shades*) in the field gradient at the electrode edge and a minimum (*dark shades*) in the centre between the four electrodes. The left imassge shows a two-dimensional finite element model of the field gradient while the right image shows a picture of live and dead yeast expressing positive DEP (live) and negative DEP (dead).

two opposite electrodes are connected to the same phase of the signal. In this case, positive DEP collects the particles at the electrode edges while negative DEP focuses particles into the centre between the electrodes. Polynomial electrodes can easily detect positive or negative DEP and are therefore ideal for cross-over measurements. Quantitative data can be measured for negative DEP by measuring the levitation height; however, for positive DEP, collection or velocity measurements are possible but less reliable. By using a four-phase signal generator, which applies a signal where the phase of an electrode is shifted by 90° compared to its neighbour, polynomial electrodes can be also used for electrorotation (43). This related technique uses the four-phase field to rotate cell. Here direction and speed of rotation are related to the properties of the cells. There are a number of examples for electrorotation in the literature. It is mainly used for characterising small numbers of cells to avoid cell–cell interactions in case the cells get too close to each other.

4.3. Three-Dimensional Electrode Systems

For automated measurement systems, three-dimensional electrode systems have been developed. The first system was reported by (44–46) and used two flat interdigitated electrodes forming opposite walls of a shallow flow cell. A laser beam was focused through the gap between the two-electrode arrays onto a photo diode. Positive DEP attracted cells from the liquid and thereby removed them from the light beam and thereby increases the signal on the photo diode. Negative DEP on the other hand repelled cells from the electrode array and causes a higher concentration of cells in the laser beam, which reduces the signal on the photo diode. Measuring the change in light absorption in the centre of the well gave for the first time an automated way to evaluate the strength and direction of the DEP force; however, the system relies on accurate alignment of the electrodes as well as the laser.

A more recent development in three-dimensional electrodes are DEP-well electrodes (**Fig. 8.6**) (10, 47–49). For the first time these electrodes offer a rapid a low cost way to quantify DEP. DEP-well technology uses a laminate of conducting and insulating films with wells of 500 μm to 1,200 μm drilled through the laminate. Inside these wells, the conducting end insulating layers form bands along the wall of the well similar to interdigitated electrodes. A transparent base is attached to contain the sample. Successive conductive layers of the laminate are connected to the two phases of an AC signal, to form a field gradient along the walls of the well moving cells by DEP. The redistribution of the cells can be monitored with a video camera connected to a microscope (with transmitted light illumination). An image of the well is captured before the field is applied and compared to an image that was taken while the electric field was applied for a defined

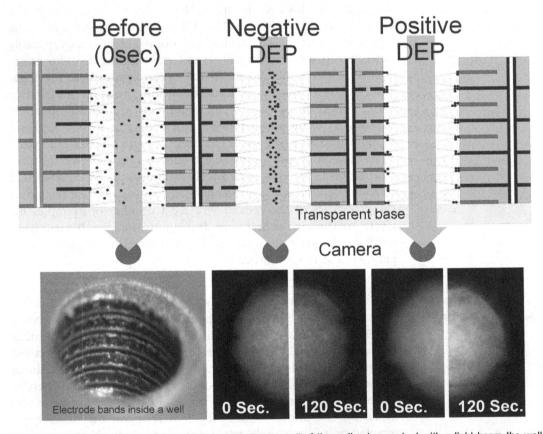

Fig. 8.6. In DEP well, positive DEP attracts the cells to the wall of the well; when probed with a light beam the well becomes more transparent. Negative DEP pushes the cells towards the centre of the well so that the well becomes less transparent.

time (e.g. 90 s). The difference in average brightness between these two images indicated the strength and direction of the DEP force. Positive DEP pulls cells towards the walls and removes them from the bulk liquid and therefore makes the well appear lighter, while negative DEP pushed cells towards the centre of the well and thereby makes the well appear darker. Since the field is applied from the walls of the well, the force is much stronger around the outside of the well. So, analysing a ring approximately half the radius of the well along the outside of the well, gives a better signal to noise ratio than analysing the whole well.

From the automation point of view DEP-well technology offers several advantages. First of all, the well (and with it the electric field) is radially symmetric which allows to reduce the mathematical complexity to a one-dimensional system. Fabrication is significantly easier since industrial lamination technologies can produce laminates with high accuracy at low cost, in particular, when compared to microfabricated electrodes. The wells also allow a better sample containment than (essentially flat) microelectrodes. This allows fabricating devices that can contain and analyse several different samples in parallel. Modern fabrication methods such as flex-PCB manufacturing also make it possible to construct devices apply different frequencies to neighbouring wells, this allows highly parallel analysis whereby several samples can be analysed at several frequencies on a single device. Devices up to 1,536 well plate standard were fabricated.

4.4. Influence of Medium

The medium used to perform DEP experiments has an important influence on their outcome since all properties are measured in relation to the medium, therefore the medium properties must be tightly controlled. Positive DEP can only be achieved when the particle is more polarisable than the medium. Media with a very high conductivity (such as most culture media) are very polarisable and therefore show only negative DEP. However, to achieve spectra with high information content, ideally, several strong transitions between positive and negative DEP are needed. Therefore most experiments are performed in media with a conductivity between 1 mS/m and 10 mS/m. As a "rule of thumb" a conductivity of 100 mS/m is the upper limit for achieving positive DEP with cells. Very low conductivity media (less than 1 mS/m) are not practical in most cases since the conductivity often changes over time, by absorbing CO_2 for the atmosphere as well as dissolving ions from surfaces or even the cells losing ions to the medium.

Lower conductivity media also have a lower current flowing through the liquid and thereby cause less joule heating. While the currents in high conductivity media can cause local "hot-spots" that can cause convention currents and electro-thermal flow (36).

The osmotic pressure of the solution is another factor to consider. Physiological solutions such as culture media contain around 280 mMol/L of dissolved particles (mostly ions) and is in osmotic equilibrium with suspended cells. If the medium would be diluted with distilled water until the conductivity is low enough for experiments, the solution would contain much less dissolved particles than the cell and thereby cause a significant osmotic pressure to the cells. To overcome this problem, non-charged molecules such as sugars are added to the solution until we have a solution that is iso-osmotic to culture medium. In general a concentration of 280 mMol/L is considered adequate. To prepare a medium, first a solution of non-charged molecules is prepared, and then a physiological salt or buffer solution (e.g. PBS, or buffers) is added until the target conductivity is reached. Common DEP media contain, e.g. mannitol 51 g/L or sucrose 85 g/L combined with glucose 3 g/L and are adjusted to the desired conductivity with KCl, culture media or pH buffers such as Phosphate buffer, PBS, and HEPES.

Before an experiment the cells are pelleted by centrifugation and the supernatant is replaced with conductivity medium. The sample is then re-suspended and centrifuged a further two times to replace all original medium. If a new cell line is investigated it is advisable to test viability of the cells after they been stored in the medium for a few hours. If the viability is reduced other conductivity mixtures should be tried. The addition of small amounts of nutrients (such as glucose) may improve the medium, while the addition of small amounts of calcium salts or EDTA may help preventing the cells clustering. However, the addition of salts changes the conductivity so it needs to be monitored.

5. Conclusion

DEP can offer a valuable tool for cell characterisation. Unlike conventional tools such as fluorescence assays, DEP is completely label-free and non-destructive. It allows investigating a rage of electrophysiological parameters in a fast and parallel assay. Since experiments can be performed using relatively simple instrumentation such as needle electrodes, it is easily accessible to a wide range of laboratories, while numerous research groups around the world develop more reliable and easier to use instrumentation and data analysis models that will develop DEP into a standard analysis technology.

References

1. Pohl, H.A. (1951) The motion and precipitation of suspensoids in divergent electric fields. *Journal of Applied Physics* 22, 155–181

2. Pohl, H.A., Hawk, I. (1966) Separation of living and dead cells by dielectrophoresis. *Science* 152, 647–649

3. Markx, G.H., Pethig, R. (1995) Dielectrophoretic separation of cells: Continuous separation. *Biotechnology and Bioengineering* **45**, 337–343

4. Gascoyne, P.R.C., Huang, Y., Pethig, R., Vykoukal, J., Becker, F.F. (1992) Dielectrophoretic separation of mammalian cells studied by computerized image analysis. *Measurement Science and Technology* **3**, 439–445

5. Medoro, G., Leonardi, A., Altomare, L., Manaresi, N., Tartagni, M., Guerrier, R. (2003) A lab-on-a-chip for cell detection and manipulation. *IEEE Sensors Journal* **3**, 317

6. Gascoyne, P.R.C., Mahidol, C., Ruchirawat, M., Satayavivad, J., Watcharasit, P., Becker, F.F. (2002) Microsample preparation by dielectrophoresis: isolation of malaria. *Lab on a Chip* **2**, 70–75

7. Gascoyne, P.R.C., Wang, Y., Huang, X.B., Becker, F.F. (1997) Dielectrophoretic separation of cancer cells from blood. *IEEE Transactions on Industry Applications* **33**, 670–678

8. Labeed, F.H., Coley, H.M., Thomas, H., Hughes, M.P. (2003) Assessment of multidrug resistance reversal using dielectrophoresis and flow cytometry. *Biophysical Journal* **85**, 2028–2034

9. Labeed, F.H., Coley, H.M., Hughes, M.P. (2006) Differences in the biophysical properties of membrane and cytoplasm of apoptotic cells revealed using dielectrophoresis. *Biochim Biophys Acta* **1760**, 922–929

10. Hubner, Y., Hoettges, K.F., Kass, G.E., Ogin, S.L., Hughes, M.P. (2005) Parallel measurements of drug actions on Erythrocytes by dielectrophoresis, using a three-dimensional electrode design. *IEEE Proceedings of Nanobiotechnology* **152**, 150–154

11. Hölzel, R. (2002) Single particle characterization and manipulation by opposite field dielectrophoresis. *Journal of Electrostatics* **56**, 435–447

12. Markx, G.H., Huang, Y., Zhou, X.F., Pethig, R. (1994) Dielectrophoretic characterization and separation of micro-organisms. *Microbiology* **140**, 585–591

13. Johari, J., Hubner, Y., Hull, J.C., Dale, J.W., Hughes, M.P. (2003) Dielectrophoretic assay of bacterial resistance to antibiotics. *Physics in Medicine and Biology* **48**, N193–N198

14. Huebner, Y., Hoettges, K.F., Hughes, M.P. (2003) Water quality test based on dielectrophoretic measurements of fresh water algae Selenastrum capricornutum. *Journal of Environmental Monitoring* **5**, 861–864

15. Hughes, M.P., Morgan, H., Rixon, F. (2002) Measuring the dielectric properties of herpes simplex virus type 1 virions with dielectrophoresis. *Biochimica et Biophysica Acta* **1571**, 1–8

16. Hughes, M.P., Morgan, H. (1999) Dielectrophoretic characterization and separation of antibody-coated submicrometer latex spheres. *Analytical Chemistry* **71**, 3441–3445

17. Hughes, M.P., Morgan, H., Flynn, M.F. (1999) The dielectrophoretic behavior of submicron latex spheres: influence of surface conductance. *Journal of Colloid and Interface Science* **220**, 454–457

18. Mureau, N., Mendoza, E., Silva, S.R.P., Hoettges, K.F., Hughes, M.P. (2006) In situ and real time determination of metallic and semiconducting single-walled carbon nanotubes in suspension via dielectrophoresis. *Applied Physics Letters* **243109**, 88–90

19. Pethig, R., Markx, G.H. (1997) Applications of dielectrophoresis in biotechnology. *Trends in Biotechnology* **15**, 426–432

20. Kua C.H., Lam, Y.C., Yang, C., Youcef-Toumi, K. (2005) Review of bio-particle manipulation using dielectrophoresis. *Innovation in Manufacturing Systems and Technology*

21. Wang, X.B., Huang, Y., Gascoyne, P.R.C., Becker, F.F. (1997) Dielectrophoretic manipulation of particles. *IEEE Transactions on Industry Applications* **33**, 660–669

22. Hughes, M.P., Morgan, H., Rixon, F.J., Burt, J.P.H, Pethig, R. (1998) Manipulation of herpes simplex virus type 1 by dielectrophoresis, *Biochim Biophys Acta* **1425**, 119–126

23. Green, N.G., Morgan, H., Milner, J.J. (1997) Manipulation and trapping of submicron bioparticles using dielectrophoresis, Journal of *Biochemical and Biophysical Methods* **35**, 89–102

24. Schnelle, T., Müller, T., Fiedler, S., Shirley, S.G., Ludwig, K., Herrmann, A., Fuhr, G., Wagner, B., Zimmermann, U. (1996) Trapping of viruses in high-frequency electric field cages. *Naturwissenschaften* **83**, 172–176

25. Huang, Y., Holzel, R., Pethig, R., Wang, X-B (1992) Differences in the AC electrodynamics of viable and non-viable yeast cells determined through combined dielectrophoresis and electrorotation studies. *Physics in Medicine and Biology* **37**, 1499–1517

26. Watarai, H., Sakomoto, T., Tsukahara, S. (1997) In situ measurement of dielectrophoretic mobility of single polystyrene microparticles. *Langmuir* **13**, 2417–2420

27. Jones, T.B., Bliss, G.W. (1977) Bubble dielectrophoresis. *Journal of Applied Physics* **48**, 1412

28. Kaler, K.V., Jones, T.B. (1990) Dielectrophoretic spectra of single cells determined by feedback-controlled levitation. *Biophysical Journal* **57**, 173–182

29. Liu, X., Spencer, J.L., Kaiser, A.B., Arnold, W.M. (2004) Manipulation and purification of multi-walled carbon nanotubes. *Annual Conference on Electrical Insulation and Dielectric Phenomena 2004. (CEIDP '04)* 336–339

30. Allsopp, D.W.E., Milner, K.R., Brown, A.P., Betts, W.B. (1999) Impedance technique for measuring dielectrophoretic collection of microbiological particles. *Journal of Physics D: Applied Physics* **32**, 1066–1074

31. Linderholm, P., Seger, U., Renaud, P. (2006) Analytical expression for electric field between two facing strip electrodes in microchannels. *Electronics Letters* **42**, 145–147

32. Irimajiri, A., Hanai, T., Inouye, A. (1979) A dielectric theory of "multi-stratified shell" model with its application to a lymphoma cell. *Journal of Theoretical Biology* **78**, 251–269

33. Jones, T.B. (1995) *Electromechanics or Particles.* Cambridge University Press, Cambridge

34. Sanchis, A., Sancho, M., Martínez, G., Sebastián, J.L., Muñoz, S. (2004) A BEM modelling of the dielectrophoretic behaviour of bacterial cells. *URSI EMTS*

35. Hughes, M.P. (2003) *Nanoelectromechanics in Engineering and Biology. Nano- and Microscience, Engineering, Technology, and Medicine.* CRC Press, Boca Raton

36. Morgan, H., Green, N.G. (2003) *AC Electrokinetics: Colloids and Nanoparticles.* Research Studies Press, Baldock, UK

37. Broche, L.M., Labeed, F.H., Hughes, M.P. (2005) Extraction of dielectric properties of multiple populations from dielectrophoretic collection spectrum data. *Physics in Medicine and Biology* **50**, 2267–2274

38. Pethig, R., Huang, Y., Wang, X.B., Burt, J.P.H. (1992) Positive and negative dielectrophoretic collection of colloidal particles using interdigitated castellated microelectrodes. *Journal of Physics D: Applied Physics* **25**, 881–888

39. Hunag, Y., Pethig, R. (1991) Electrode design for negative dielectrophoresis application. *Measurement Science and Technology* **2**, 1142–1146

40. Wang, X.B., Huang, Y., Burt, J.P.H., Markx, G.H., Pethig, R. (1993) Selective dielectrophoretic confinement of bioparticles in potential wells. *Journal of Physics D: Applied Physics* **26**, 1278–1285

41. Markx, G.H., Huang, Y., Zhou, X.F., Pethig, R. (1994) Dielectrophoretic characterization and separation of micro-organizms. *Microbiology* **140**, 585–591

42. Pohl, H.A. (1978) *Dielectrophoresis.* Cambridge University Press, Cambridge, UK

43. Wang, X.B., Huang, Y., Holzel, R., Burt, J.P.H., Pethig, R. (1993) Theoretical and experimental investigations of the interdependence of the dielectric, dielectrophoretic and electrorotational behaviour of colloidal particles. *Journal of Physics D: Applied Physics* **26**, 312–322

44. Price, J.A.R., Burt, J.P.H., Pethig, R. (1988) Applications of a new optical technique for measuring the dielectrophoretic behaviour of micro-organisms. *Biochimica et Biophysica Acta (BBA) – General Subjects* **964**, 221–230

45. Burt, J.P.H., Al-Ameen, T.A.K., Pethig, R. (1989) An optical dielectrophoresis spectrometer for low-frequency measurements on colloidal suspensions. *Journal of Physics E: Scientific Instruments* **22**, 952–957

46. Talary, M.S., Pethig, R. (1994) Optical technique for measuring the positive and negative dielectrophoretic behaviour of cells and colloidal suspensions. *IEE Proceedings – Science, Measurement and Technology* **141**, 395–399

47. Fatoyinbo, H.O., Kamchis, D., Whattingham, R., Ogin, S.L., Hughes, M.P. (2005) A high-throughput 3-D composite dielectrophoretic separator. *IEEE Transactions on Biomedical Engineering* **52**, 1347–1349

48. Hughes, M.P., Hoettges, K.F. (2005) Dielectrophoresis for drug discovery and cell analysis: novel electrodes for high-throughput screening. *Biophysical Journal* **88**, 172A–172A

49. Hoettges, K.F., Hübner, Y., Kass, G.E.N., Ogin, S.L., Hughes, M.P. (2005) Dielectrophoresis as a screening tool – label-free electrophysiology assays. *Screening – Trends in Drug Discovery* **3/2005**, 28–20

Chapter 9

AC-Electrokinetic Applications in a Biological Setting

Fatima H. Labeed

Abstract

Dielectrophoresis is a phenomenon which can be exploited to provide significant quantitative electrophysiological data in a range of biochemical setting, from oncology to drug discovery. This chapter seeks to elucidate those applications and the electrophysiological phenomena underpinning those applications.

Key words: AC-electrokinetics, dielectrophoresis, cancer, drug, cell.

1. Introduction

AC-electrokinetics is the collective term used to describe a group of phenomena such as dielectrophoresis (DEP) and electrorotation (ROT). The motion of suspensoid particles resulting from polarisation forces produced by an inhomogeneous electric field was first described and termed *dielectrophoresis* by Herbert Pohl (1950) (1). He determined that this phenomenon was related to the related phenomenon of electrophoresis, in which suspended particles are moved by the action of an electrostatic field on the charged particles.

If a dielectric particle (such as a cell) is suspended in an electric field, it will polarise and acquire a dipole. The direction and magnitude of this induced dipole depends on the frequency and magnitude of the electric field and on the dielectric properties of the cell and the medium (2). Historically, the use of non-uniform electric fields was suggested by Pohl to have dated back to the nineteenth century. The DEP phenomenon was widely applied, later on, to the manipulation, separation and analysis of cellular and viral particles (3–5).

M.P. Hughes, K.F. Hoettges (eds.), *Microengineering in Biotechnology,* Methods in Molecular Biology 583,
DOI 10.1007/978-1-60327-106-6_9, © Humana Press, a part of Springer Science+Business Media, LLC 2010

Particles experiencing such forces can be made to exhibit a variety of motions including attraction to, and repulsion from, regions of high electric field (termed positive and negative DEP, respectively) by changing the frequency of the applied electric field. Pohl (2) described how bio-particles such as bacteria, cells or viruses can be characterised by their dielectric properties, and since then dielectrophoresis has been used to distinguish between different types of bacteria (6), to detect changes in cell cytoplasmic properties (7) and to detect whether cells are viable or non-viable (8, 9). The main factors influencing the dielectric properties of a bio-particle are the surface charge, the membrane capacitance and the conductivity of the cytoplasm. If drugs such as antibiotics change any of these factors, the dielectric properties of the cell change and can be detected (10); this change can then be used to distinguish between cells that are resistant or sensitive to a drug.

By exploiting the fact that different particles may experience forces acting in different directions when all other factors are the same, researchers have been able to use DEP to analyse and separate mixtures of cells on electrode arrays. For example, work has demonstrated that DEP can be used to examine the effect of drugs on cells, such as the effect of nystatin on erythrocytes (11) or the response of neutrophils to activation by chemotactic factors (12). Since populations of particles may experience forces acting in different directions within specific frequency windows, separation has been demonstrated for mixtures of viable and non-viable yeast cells (13), and CD34+ cells from bone marrow (14), which can be separated from human blood. Other demonstrations of cell separation have been made in separating breast cancer cells from blood, erythrocytes with and without malarial parasite infection, leukocytes, different types of viruses in solution and, more recently, white blood cells form erythrocytes using this technique (15–18).

Dielectrophoresis can be used to determine the dielectric properties of particles by examining the behaviour of particles across a broad frequency range. This typically involved the determination of a frequency where the dielectrophoretic force on a particle is zero, and hence determining the dielectric properties of the membrane (19). This method has been applied to studies of cellular responses to a broad range of toxicants (20), membrane changes during apoptosis (21), malarial infection (22) and cancer cell transformation (23). Studies of the dielectrophoretic behaviour of cells over a broader frequency range have elicited information about both membrane and cytoplasm, and have been used to examine the effects of antibiotics on bacteria (10), the effects of copper sulphate on algae (24). This potentially makes DEP a very valuable tool for screening applications.

To extract dielectric parameters from dielectrophoretic data, modelling techniques are used to find best-fit parameter for given sets of data indicating the time-dependent polarisibility of the

particles. Analytical expressions have been derived for the dielectrophoretic behaviour of homogeneous spheres (25). However, it is
generally considered (3) that the expansion of the equation linking
the dielectrophoretic behaviour to the dielectric properties of the
particle is too complex to provide useful analytical expressions directly
linking the dielectric properties of the cell to the observed dielectrophoretic behaviour. In general, the estimation of dielectric properties
is performed by best-fit numerical analysis (23) in order to determine
the properties of the membrane and the cytoplasm to provide information of potential significance to biological scientists.

2. Theory

A polarisable particle suspended in a uniform electric field will
experience no net movement, as the forces induced by the interaction between each of the dipolar charges and the electric field are
equal and opposite; the cell will not move unless it carries a net
charge. Even when there is a net charge, the particle will not
undergo observable displacement unless the field frequency is
equal to, or near, zero.

The interaction of the charges on each side of the dipole with the
electric field generates a force. Coulomb's law states that a charge Q_1
generates an electric field E and that this electric field induces a force
on second charge Q_2 according to the following expression [1]:

$$E = \frac{Q_1}{4\pi\varepsilon d^2}r$$

$$F = Q_2 E$$

[1]

where d the distance between the charges, r is the unit vector
directed from Q_1 to Q_2 and ε is the permittivity of the material
surrounding Q_2.

If there is an electric field gradient across the cell, the magnitude of the forces on either side of the cell will be different. The cell
will thus experience a net movement in the direction of increasing
field strength.

The particle will always move along the direction of greatest
increasing electric field, regardless of the field polarity; the dipole
will re-orient itself with the applied polarity (AC or DC fields), and
the force is always governed by the field gradient rather than the
field orientation. This force is termed dielectrophoresis (DEP).

Dielectrophoresis can be further classified by direction (3). If
the particle is more polarisable than the surrounding medium, it
will experience a phenomenon called *positive DEP*, where cells are

attracted to the electrodes. Its opposite effect, where cells are repelled from the electrodes, is called *negative DEP*. Whether a particle experiences positive or negative DEP is governed by its polarisibility relative to the suspending medium. It is the positive dielectrophoresis phenomenon that is used in this work.

The dielectrophoretic force, F_{DEP}, acting on a spherical body can be used to obtain parameters such as permittivity and conductivity. The time-averaged force exerted on a spherical particle of radius r suspended in a medium of relative permittivity ε_r and exposed to an electric field gradient is given by equation [2]:

$$F_{DEP} = 2\pi\varepsilon_0\varepsilon_r r^3 \text{Re}[K(\omega)]\nabla E^2, \qquad [2]$$

where ∇E^2 is the gradient of the strength of the applied electric field squared and Re $[K(\omega)]$ is the real part of the Clausius–Mossotti factor given by equation [3]:

$$K(\omega) = \left(\frac{\varepsilon_p^* - \varepsilon_m^*}{\varepsilon_p^* + 2\varepsilon_m^*}\right), \qquad [3]$$

where ε_p and ε_m are the complex permittivities of particle and medium respectively, given by equation [4]:

$$\varepsilon_i^* = \varepsilon_i - j\frac{\sigma_i}{\omega}, \qquad [4]$$

where ε_i and σ_i refer to the (real) permittivity and conductivity of material i.

Characterisation of cells is made on the basis of physical phenomena occurring inside the cell, such as changes in conductivity (which reflects the ability to carry the electric charge) and permittivity (reflecting the ability to store the electric charge), for both cytoplasm and the membrane. Where the particle, rather than being homogeneous, consists of a shell surrounding a homogeneous core. As described by Huang (8), the smeared-out sphere approach can be used to determine the effective frequency-dependent complex permittivity of multi-shelled particles. We can replace the complex permittivity with an effective value combining the properties of the shell and the core, as follows:

$$\varepsilon_{1eff}^* = \varepsilon_2^* \frac{\left(\frac{r_2}{r_1}\right)^3 + 2\frac{\varepsilon_1^* - \varepsilon_2^*}{\varepsilon_1^* + 2\varepsilon_2^*}}{\left(\frac{r_2}{r_1}\right)^3 - \frac{\varepsilon_1^* - \varepsilon_2^*}{\varepsilon_1^* + 2\varepsilon_2^*}}, \qquad [5]$$

where subscripts 1 and 2 correspond to the core and shell, and r_1 and r_2 are the radii from the centre of the sphere to the inside and outside of the membrane so that the Clausius–Mossotti factor for the ensemble is given by equation [6]:

$$K(\omega) = \frac{\varepsilon_{1\text{eff}}^* - \varepsilon_3^*}{\varepsilon_{1\text{eff}}^* + 2\varepsilon_3^*}, \qquad [6]$$

where the subscript 3 refers to the suspending medium. If the full expression for $\text{Re}[K(\omega)]$ is expressed analytically, it is sufficiently complex to render direct mathematical analysis hopelessly complicated (3); instead, researchers have used numerical methods to determine dielectric properties by analysis of dielectrophoretic spectra. Dielectrophoretic data have been analysed (15) by simplifying the solution using assumptions and by examining a number of crossover spectra as a function of medium conductivity. However, such methods are limited, particularly where the sample of cells precludes the repeating of sample analysis or where multiple populations are present.

A given sphere, where the shell is of relatively low conductivity compared to the inner compartment and outer medium, has a polarisibility spectrum which displays two characteristic dispersions, one rising at lower frequency and one falling at higher frequency. The frequency where the polarisibility crosses from negative to positive for a homogeneous sphere allows the direct determination of the properties of the sphere from equation [3] by equating it to zero; however, for a shelled sphere the expression is much more complicated. Some work has been performed in deriving expressions for the lower crossover frequency, but the upper frequency has largely been ignored because of limitations with signal generation equipment, as the upper crossover frequency is typically up to an order of magnitude higher than the maximum frequency of most signal generators.

Benguigui and Lin (25) demonstrated that it is possible to determine an expression for the low frequency crossover of a homogeneous sphere of conductivity and permittivity ε_p and σ_p, respectively; this was extended by Huang et al. (23) to consider the low-frequency crossover of a shelled sphere by considering it to be equivalent to a homogeneous sphere of effective conductivity and permittivity of

$$\left.\begin{array}{l} \varepsilon_p = \varepsilon_2 \left(\dfrac{r2}{r2 - r1} \right) \\[3mm] \sigma_p = \sigma_2 \left(\dfrac{r2}{r2 - r1} \right) \end{array}\right\} . \qquad [7]$$

Using this approximation, Huang and co-workers produced an expression for the low-frequency crossover F_{x1}:

$$F_{x1} = \frac{1}{2\pi} \sqrt{\frac{2\sigma_3^2 - A\sigma_2\sigma_3 - A^2\sigma_{23}^2}{A^2\varepsilon_2^2 - A\varepsilon_2\varepsilon_3 - 2\varepsilon_3^2}}, \qquad [8]$$

where $A = \left(\frac{r_2}{r_2 - r_1}\right)$, that is, the ratio of the radius of the cell to the thickness of the membrane.

At low frequencies, as reviewed (26), the total current comprises a surface current around the cell, plus a bulk membrane current, both having radial and tangential components. The overall conductivity of the particle has an additional term given by

$$\sigma_p = \sigma_{p\text{bulk}} + \frac{2K_s}{r}, \qquad [9]$$

where $\sigma_{p\text{bulk}}$ is the bulk conductivity of the cell membrane, K_s is the surface conductance and r is the cell radius.

It is known that for the majority of cells, the thickness of the membrane is considerably smaller than the radius of the cell, that is, the ratio $a = \left(\frac{r_2}{r_1}\right)^3$ has a value very near 1. If we make the approximation $a = 1$, we find that the high-frequency crossover frequency F_{x2} is given by the following:

$$F_{x2} = \frac{1}{2\pi} \sqrt{\frac{\sigma_1^2 - \sigma_1\sigma_3 - 2\sigma_3^2}{2\varepsilon_3^2 - \varepsilon_1\varepsilon_3 - \varepsilon_1^2}}. \qquad [10]$$

Numerical studies of the output of the full expression indicate that this approximation holds for the upper crossover frequency for a wide range of electrical values provided $a < 1.15$ (approx.), corresponding to a ratio of cell radius to membrane thickness of 20:1. Within this limit, which represents all cases of biological cells, equation [10] holds. The only case where the equation does not work is where no upper crossover exists, which happens when $\varepsilon_1 > \varepsilon_3$ (meaning the real part of the Clausius–Mossotti factor remains positive for all frequencies above F_{x1}) or $\sigma_1\sigma_3$ (meaning the real part of the Clausius–Mossotti factor never has a value greater than zero). The former is a highly unlikely event in biology and the latter can be controlled by the use of low-conductivity media. Provided these conditions are not met, ε_1 and σ_3 have very little effect on F_{x2}. Since the upper crossover frequency is independent of size, largely independent (in biological cells) of cytoplasmic permittivity or membrane properties, and of the remaining variables those relating to the medium can be precisely defined, F_{x2} provides a direct measurement of the cytoplasmic conductivity.

This result is significant for determining the upper dispersion characteristics. Although the actual crossover frequency is often too high to be observed with conventional function generators, the behaviour of the dielectric dispersion – taking approximately one decade to transit between stable plateaux of $Re[K(\omega)]$ – means that the frequency at which $Re[K(\omega)]$ begins to decline is *also* dependent only on σ_1. This can be extended to multiple populations, as described in (27); the presence of multiple upwards- or downwards-pointing dispersions can be related to the presence of multiple populations, where the downwards-pointing dispersions (with increasing frequency) points to populations categorised solely by their cytoplasmic conductivities.

3. DEP for Biological Characterisation

The first application of DEP to living cells was described by Pohl and Hawk (28), where they described the first means of physically separating live and dead cells. Since then, DEP and related phenomena have been shown to be useful in a variety of biological systems, including algae (24), bacteria (6), yeasts (8, 13) and mammalian white blood cells (16) and red blood cells (29). The use of DEP in treating aqueous suspensions of cells or organelles is generally restricted to using *alternating* electric (AC) fields rather than *static* fields as applied to electrophoresis, which is a major point of difference in the two techniques (2); this is because in AC fields the average displacement due to the electrophoretic force component is zero.

Without a model of the behaviour of complex bio-particles such as cells, Pohl (2) interpreted positive DEP behaviour in the light of known biophysical properties of cells. Working on the assumption that positive DEP was driven entirely by a particle having greater permittivity than the medium, he described how water is a high polar material, and as such will be pulled strongly towards the region of highest field intensity by the non-uniform field. If the cell was to move to the region of highest field intensity, it must therefore exhibit an even higher specific polarisibility. Pohl (2) described a number of ways the cellular system can attain this high polarisibility. First, the cell itself is largely water; second, there are dissolved intracellular regions of numerous polar molecules-proteins, sugars, DNA and RNA, all of which can contribute to polarisation. Third, there are structured regions that can act as capacitive regions, e.g. lipid membranes across which electrolytes can act to produce charge distributions. Fourth, there are structured areas in the surface where ionic double layers can produce very large polarisations. As frequency increases, the conductivity of membrane and cytoplasm also make important contributions. Since that time, it has been shown that Maxwell-Wagner interfacial polarisation is the dominant source of dipolar polarisation, meaning that conductivity and surface charge can also play a significant role in determining polarisibility, e.g. see (5) for explanation.

The first successful experiments reported of the use of DEP to collect living cells were by Pohl and Hawk (28). The authors used very simple equipment – a voltage supply, a pair of small, bare wire (pin), plate electrodes and a microscope. They showed that yeast (*S. cerevisiae*) could be made to collect at a reliable rate in the region of highest field intensity. Other yeast work was carried out using different configurations, such as an arrangement of two opposing pins.

In the early period of DEP study, no adequate model existed for explaining frequency-dependent DEP behaviour of complex bio-particles, so empirical studies were performed using DEP and yeast cells by varying the experimental conditions and observing the effect on the number of cells collected. This was measured as the average length of the pearl chain groups of cells attached to the electrodes after a particular time of collection. From this, the dielectrophoresis collection rate (DCR) was obtained, defined as the yield per unit time. The units of yield were divisions of the reticule of a microscope eyepiece (1 division corresponding to an object length of 10.3 μm). Physical parameters varied included voltage, cell concentration and frequency and conductivity (2). The response of the cells was found to be approximately linear with that of voltage, but at much higher voltages, the yield was reduced. This reduction was suggested to be primarily due to the "stirring" effect (now called electrothermal flow), which results from the application of intense fields in free liquids. When the stirring is moderate it brings in more cells close to the pins where they can be more readily collected by the strong field gradient on the pins. At higher voltages, this effect becomes so strong that the flow becomes turbulent and prevents cells from reaching the electrodes, hence reducing the yield. The yield was also found to be linear with the cell concentration. The overall results (2) concluded that voltage, concentration and time elapsed affect the yield more or less independently of other variables, whereas frequency and conductivity were strongly coupled, where one must be specified before the effect of the other can be given.

Other experiments with yeast cells involved investigating the effects of external agents on collection of cells. The biological parameters investigated included colony age, heat treatment, exposure to UV radiation and treatment with herbicides. Pohl and Hawk (28) noticed that the organisms were greatly affected by heat and demonstrated that heat-killed (or autoclaved) yeast cells could be physically separated from the living by DEP. DEP yield of yeast cells that had been irradiated with UV light has been studied. After treating the cells at a wavelength that was believed to have inflicted nuclear damage, DEP results showed that selective UV treatment did not affect the polarisation responses of cells. DEP has also been exploited in the study of herbicides to determine the effectiveness of the herbicides tested (as reviewed in (2)).

To examine the effect of ageing yeast colonies (*S. cerevisae*), samples from different incubation periods were compared (30) using DEP over a range of frequencies (10^2–10^7Hz). When relatively high solution conductivity was used, the very old cells had a slightly higher yield at low frequencies and a slightly lower yield at high frequencies. The younger cells differed from the older cells by exhibiting no collection at 100 Hz.

Further, early DEP studies included comparing canine blood platelets, or thrombocytes from normal male and female dogs, and their Factor VIII deficient counterparts, reviewed in (2). The yield DEP spectra were compared, and moderate differences in the polarisability of normal and various haemophilic type blood platelets were observed, and bacteria such as *flavobacterium*, and *E. coli*, *P. aeruginosa*, *B. megaterium* and *B. cereus*.

In the 1980s, DEP research concentrated primarily on two aspects: the search for a model capable of accurately predicting DEP response and further exploitation of the technique for biological applications. One such application was the use of DEP to bring cells together for electrofusion, reviewed in (4). Other applications, pursued largely by Dimitrov and colleagues, saw DEP applied to characterisation of red blood cells (31), a range of cell types myelomas, hybridomas and lymphocytes B (32), and white blood cells (33), among others. During this period, the discovery of electrorotation by Arnold and Zimmermann (1982) (34) allowed a second electrokinetic method of cell analysis; this was applied for the measurement of membrane capacitance of single mesophyll cells in avena-sativa (oats), and the rotation of a single cell in a rotating field (34, 35). Subsequent work used ROT to determine the dielectric properties of shelled spheres such as cells using theory developed by Fuhr and others, reviewed in (36).

In the 1990s, DEP research expanded dramatically due to a number of factors. First, the *"smeared-out"* multi-shell model (originally developed by Hanai (37), then by Irimajiri and colleagues (38), and later applied by Huang and colleagues (8)) supplanted earlier methods of determination of electrical properties, allowing DEP and ROT data to be interpreted more readily. For example, Chan and colleagues (39) used multi-shell models to describe the electrokinetic behaviour of liposomes, and they demonstrated that the information provided by DEP and ROT yielded values of dielectric parameters of liposome-like particles. Second, application of microengineering to the production of microelectrodes allows easy and direct observation of cell DEP in any laboratory. Finally, the growth of the "lab-on-a-chip" movement brought an increase in interest in microengineered analysis and separation devices, particularly those with potential commercial applications.

Positive and negative DEP were first demonstrated on microelectrodes in 1992 by Pethig and colleagues (40) and Gascoyne and colleagues (41). The former group used yeast cells to show that collection arising from both positive and negative DEP was facilitated using the castellated electrode geometry, and the two forms of collection differed significantly as functions of the frequency of the applied non-uniform electric field and of the conductivity of the suspending medium. The latter group (41) used positive and negative DEP to characterise normal, leukaemic and differentiation-induced leukaemic mouse erythrocytes as a function of frequency in the range 5×10^2–10^5Hz, which were shown to be significantly different. The group also demonstrated that by choosing the suitable frequency and cell suspension medium, DEP could be used to separate these cells.

Dielectrophoresis was also used as a characterisation technique where the dielectric properties were used to provide the basis of subsequent separation work. For example, the effective electrical conductivity values for Gram-positive and Gram-negative bacteria were determined in the frequency range (10–100 kHz) (6), where this information enabled experimental conditions to be selected to separate different microorganism species using DEP.

Another example where DEP and ROT were used in combination was that by Huang and colleagues (1996) (23), as the two techniques were used to determine the dielectric properties of a clone of normal rat kidney cells, which exhibited a transformed phenotype at 33°C and a non-transformed phenotype at 39°C. DEP measurements of the crossover frequency (at which cells experienced zero force) as a function of the suspension medium revealed that, in response to a temperature shift from 33°C to 39°C for 24 h, a significant decrease was observed in the specific membrane capacitance and conductance. ROT analyses demonstrated a similar reduction in the membrane capacitance but showed insignificant changes in the internal conductivity. The changes observed in the membrane dielectric properties and the membrane specific capacitance was correlated with cell membrane surface or morphology complexity, as confirmed by scanning electron microscopy. Other studies have considered the assessment of biocide action on bacterial biofilms (42), for the detection of toxic chemicals.

The use of DEP towards the mid and late 1990s showed an increased interest in the microbiology field. Quinn and colleagues (43) investigated this technology for the rapid analysis of ozonated *Cryptosporidium parvum* oocysts, where two key frequencies were used for DEP collection (100 kHz and 10 MHz) to compare the effects of ozonation at various dosages. A consistent pattern was suggested between increasing the ozone dose and a decrease in oocyte internal conductivity, as supported using a multi-shell mathematical modelling of spheres. The use of DEP as part of microbial analysis included characterisation of microbes in water (44), where the cell collection spectra were determined and used as the basis for distinguishing different cell types. This analytical application was then suggested as a method of separation in microbiology. Water quality testing and healthcare in general have provided ideas for exploiting DEP. One study (45) investigated the electrical properties of two waterborne protozoan parasites, *Giardia muris* and *Cyclospora cayetanesis*, with the potential use of the technique as a rapid and a precise assay for the determination of parasite viability. A more recent study (24) suggested the potential use of DEP as a low-cost and a rapid method of testing water quality, as the authors used DEP for the dielectrophoretic measurements of fresh water algae *Selenastrum capricornutum* in the presence of different copper sulphate concentrations. A more recent application of DEP and the environment was the study by Fatoyinbo and colleagues (2006) (46), in which the group describe the separation of *Bacillus subtilis* spores from environmental diesel particles. The results highlighted the 99% successful removal of diesel particulates acquired from environmental samples, whilst allowing the bacterial spores pass through the chamber unimpeded. The results of this study demonstrated the potential of the DEP device to act as a precursor to a range of detection methods including those of biowarfare.

One limitation of DEP for automated study is the requirement to observe cell motion. In the late 1990s, Suehiro and colleagues (47) described a new technique that realises the estimation of biological cell concentration in an aqueous medium. The group used a dielectrophoretic impedance measurement (DEPIM) method between two electrodes, and used positive DEP to capture cells on an interdigitated electrode system in pearl-chain formation. By monitoring the variations in the electrical impedance, cell populations were analysed quantitatively according to theoretical model of cell collection process, where a suspension of *E. coli* could accurately be assayed in about 10 min at $10^5/cm^{-3}$ concentration. In a more recent literature source (48), a newer method was proposed to improve DEPIM sensitivity using an electropermeabilisation technique, which was applied after using DEP to trap cells in order to release intracellular ions. When high AC fields were applied across the trapped bacteria, an increase in conductance was observed, which was suggested to be due to electropermeabilisation.

Since DEP allows the determination of drug action, it has potential for routine testing of the effects of drugs for pharmaceutical applications. A study (10) described the application of DEP, to determine drug resistance by studying the dielectric parameters of *Staphylococcus epidermidis*. The results indicated a strong similarity between the dielectric properties of sensitive and resistant strains in the absence of antibiotic treatment, whereas there was a significant difference between the sensitive and resistant strains after treatment. A more recent study demonstrated the expansion of DEP applications to include the food sector. Ali and Bashier (49) investigated the effect of fast green dye (FG), a widely used food colourant by the cosmetic and drug industry, on the biophysical properties of thymocytes and splenocytes in albino mice. Using a frequency range 20–100,000 Hz, their estimated dielectric parameters indicated that FG is an immunotoxic agent.

Owing to the fact that DEP allows the non-invasive study of cells, the obvious applications were in the analysis of human cells for potential medical applications. A number of approaches have been taken. One of which was the challenge of using DEP for the production of small and low cost automated devices for assessing parasite concentrations with potential drug sensitivity studies and diagnosis of malaria (22). The other and most common approach was that of studying cancer and it is discussed in more detail later.

Since DEP is most readily applied to suspended cells, the majority of work has been performed on blood cells. To this end, studies have been performed on the most abundant cells in the blood, including erythrocytes (50), lymphocytes (51), leukocytes (16) and platelets (52, 53). Furthermore, Chan and co-workers (54) conducted a study of circulating trophoblast cells, which detach from the placental lining and enter the circulation during pregnancy, as a potential means of isolating and testing foetal DNA.

4. Cancer Studies Using DEP

Malignant cancers are a major cause of morbidity in the UK, causing 153,491 deaths in 2005 (Cancer Research, UK). It is the subject of extensive research across many disciplines. Chemotherapy is considered to be the most effective treatment for metastatic tumours. However, cancer cells can become simultaneously resistant to different chemotherapeutic drugs that have no common structure, a phenomenon called multidrug resistance (MDR). Although a broad spectrum of treatments have been developed, understanding the behaviour of cancer cells and the determination of the prognosis and early detection is important; this is where DEP has potential applications. AC-electrokinetic techniques have been amongst the methods used to investigate cancer cells by ways of characterising them relative to their non-cancerous counterparts. Changes in the dielectric properties could potentially be used to separate cancer cells from blood. In this section, we will review how DEP and AC-electrokinetics have been used to characterise cancer cells, assess their behaviour before and after treatment and form the basis of techniques to separate cancerous from non-cancerous cells. The cellular biophysical properties can reflect cellular states such as differentiation and viability (54, 19) and DEP-based manipulation of tumour cells have formed part of the lab-on-a-chip based devices in cancer research, reviewed by Gambari and colleagues (55).

The use of DEP for characterisation of cancer cells dates back to the 1980s, where it was used to characterise human malignant melanocytes (56), myeloma and hybridoma cells and lymphocytes (33). Cristofanilli and co-workers (57) used automated ROT to examine dielectric variations related to HER-2/neu overexpression in breast cancer cell sublines (MCF-7). The study evaluated the behaviour of MCF-7/neo and p185(neu) transfectants (MCF/HER2-11 and MCF/HER2-18) to investigate whether differences in HER-2/neu expression were associated with differences in dielectric properties. Western blots were used to assess the overexpression and the specific membrane capacitance was measured. The results suggested that ROT was sufficiently sensitive to detect variations in the dielectric properties in breast cancer cell lines overexpressing p185(neu), with the mean specific membrane capacitance being significantly different in MCF/neo from the other transfected sublines (MCF/HER2-11 and MCF/HER2-18). These differences were suggested to be related to the morphological alterations determined by HER-2/*neu* overexpression.

Many characterisation experiments have involved examining changes caused by treatment with different agents to assess the effect of the agent on these dielectric properties of cancer cells. An example of this was performed by Burt and colleagues (1990) (19), who used DEP to study the membrane changes accompanying the induced differentiation of Friend murine erythroleukaemia cell lines (DS19 and R1) following treatment with hexamethylene bisacetamide (HMBA) and dimethyl sulfoxide (DMSO). These are agents that induce terminal differentiation in DS19, but not in R1. The authors found that the membrane capacitance of DS19 was decreased by 30%, and the membrane conductivity to fall by a factor of five after treating DS19. No response was seen by R1 after treatment. They also found that the theoretical model was useful for comparing differences in the data, but also reported several significant discrepancies between predications and the experimental data observed, and they suggested the discrepancies to be accounted by factors such as surface charge effects and intracellular compartments. The same erythropoetic differentiation agent (HMBA) was later used (58) to investigate alterations in the plasma membrane of DS19 using ROT. Scanning and transmission electron microscopy revealed that the high membrane capacitance obtained for DS19 before treatment reflected the large area of plasma membrane associated with complex surface morphology, such as the presence of microvilli. Furthermore, ROT demonstrated that treatment with HMBA for 3 days resulted in a fall in the membrane capacitance, which correlated with a reduction in the density of microvilli. This work demonstrated that cells exposed to 72 h of differentiation had an enhanced resilience relative to their untreated counterparts, and the authors suggested that this evidences the early stages of development of the membrane skeleton which becomes fully developed in mature erythrocytes. They highlighted the value of ROT as a non-invasive method for the characterisation of viable leukaemic cells and their responses to stimuli and showed that membrane capacitance reflects membrane morphology.

The concept of MDR has been investigated by Labeed and co-workers (2003, 2007) (59, 60). The earlier study looked at the dielectric behaviour of drug sensitive K562 (chronic myelogenous leukaemia) and its MDR (K562AR = Adriamycin/doxorubicin resistant) counterpart.

Using DEP for cellular characterisation has reached to explore cancer and the process of apoptosis (programmed cell death). A study by Wang and co-workers (2002) (21) suggested the use of DEP as a more sensitive technique to indicate membrane dielectric changes, in HL60 induced apoptosis, than the conventional flow cytometric analysis of surface phosphatidylserine expression or DNA fragmentation. The membrane capacitance was measured following 1–4 h of incubations with genistein (GEN), an apoptosis-inducing agent. DEP was used in conjunction with annexin V assay and the results showed a decrease in the membrane capacitance as the incubation period with GEN increased to 4 h. Furthermore, the results were suggested to indicate that it might be possible to correlate changes in cell dielectric properties with very early stages of apoptosis and cell DEP characteristics as early and sensitive prognostic markers of apoptosis. The authors pointed to the applicability of DEP-based technology to rapidly detect, separate and quantify normal, apoptotic and necrotic cells from cell mixtures.

The DEP work on apoptosis continued, and the work by Labeed and co-workers (2006) (61) investigated the dielectric properties of the membrane and cytoplasm of human chronic myelogenous leukaemic (K562) cells post apoptotic induction, using staurosporine. The results indicated that the cytoplasmic conductivity increased markedly within the first 4 h following treatment that lasted beyond 12 h, whilst cell shrinkage increased the specific membrane capacitance. Furthermore, multiple subpopulations were detected after 24 and 48 h post treatment, and in turn indicating the dielectric changes that the cell undergoes before death. These results were compared to other conventional apoptosis monitoring techniques, such as Annexin V and TMRE (tetramethylrhodamine ethylester), and ion efflux could be inferred in the progress of apoptosis. The same group (62) demonstrated that DEP could be used as a far more rapid apoptosis detection method than existing biochemical marker methods. The results of this work showed that K562 continued to have a persistent elevation in the cytoplasmic conductivity, and occurring as early as 30 min following treatment with staurosporine. When the DEP results were compared with that of Annexin V assay, DEP was found to be a more powerful and a more informative tool, as it was capable of detecting cell electrical changes associated with very early stage apoptosis, and as early as 30 min post exposure.

Another study (20) employed DEP for the detection of cellular responses to toxicants. Paraquat, styrene oxide (SO), N-nitrosos-N-methylurea (NMU) and puromycin were used and chosen because of their different mechanisms of action, namely membrane free radical attack, simultaneous membrane and nucleic acid attack, nucleic acid alkylation and protein synthesis inhibition, respectively. For all treatments, membrane capacitance of HL60 cells decreased. The DEP responses correlated sensitively with alterations in cell surface morphology, especially microvilli and blebs, which were also observed by scanning electron microscopy. The study pointed to the DEP-based method being more sensitive to agents that had a direct action on the membrane than those agents for which membrane alterations were secondary. Direct detection of DEP responses were found to be 15 and 30 min after exposure to agents that directly damaged membrane and those acted on the cell intracellular targets.

Huang and colleagues (63) used ROT measurements to study the cytoplasmic dielectric properties of DS19 cells following HMBA treatment. After treatment, the DS19 cells exhibited a slight increment in the average interior permittivity and a decrement in the interior conductivity. Of significance was the average permittivity of cell interiors, which was larger than that of pure water ($75\ \varepsilon_0$). Using numerical simulations, the nuclear dielectric parameters were looked at to investigate the effects of nuclei on ROT spectra of intact cells, and the analysis subsequently considered other intracellular organelles such as mitochondria. The authors concluded the large permittivity observed did not result from cell nuclei or mitochondria, and instead may have arisen from the combined effects of cytoplasmic organelles. Finally, in order to assess whether DEP affects the cells exposed to non-uniform fields, Wang and Gascoyne (64) used dielectrophoretic manipulations to investigate DS19 (murine erythroleukaemia). It was found that hydrogen peroxide was produced when sugar containing media were exposed to fields outside the typical field strengths in AC-electrokinetics. The AC typical fields used for the dielectrophoretic manipulation and sorting of cells do not damage DS19 cells.

5. Separation of Cancer Cells

The use of DEP as a separation technique in a cancer setting goes back to the early 1990s. Gascoyne and co-workers (41) demonstrated the first use of DEP as a separation technique for cancer cells. The dielectrophoretic characteristics of normal, leukaemic and differentiation-induced leukaemic mouse erythrocytes were measured as a function of frequency and were shown to be significantly different; by selecting the appropriate frequency they were able to demonstrate positive and negative DEP of healthy and Friend murine erythroleukaemic cells. Since then, the majority of separators have been flow separators, either separating by flowing cells across an electrode array and trapping by positive DEP or by using field-flow fractionation (FFF) (65), where cells are repelled and are sorted according to the height they are repelled to (and hence the speed at which they flow).

The most common form of flow separation is that of using positive and negative DEP. In 1995, the first work on separation of human cancer cells was published using this method. At the time, a considerable biomedical and clinical interest in isolating cell subpopulation of bone marrow (BM) and peripheral blood stem cell collections (PBSC) emerged. These collections contain haemopoietic stem cells required for bone marrow reconstitution. The CD34+ antigen was used as a marker to identify haemopoietic precursor cells, which were strongly expressed in the primitive cells, but lost in differentiated cells. The limitation of detecting this antigen is that it is expressed on 1–4% of BM and PBSC. DEP investigations were carried out during this time period to study the dielectrophoretic separation and enrichment of CD34+ subpopulation from BM and PBSC, e.g. (66). The results indicated that maximum enrichment of stem cells (of 4.9-fold) occurred at a frequency of around 5 kHz. A follow-up CD34+ enrichment study (14) used DEP for isolating CD34+ cell populations from leukaemic patients undergoing peripheral blood stem cell harvests. The study pointed to the limitations of the conventional techniques that rely on the use of antibodies specific to the surface marker CD34+ as being expensive and time consuming, and therefore used DEP as a tool to separate CD34+ cells, as it does not rely on cell-specific markers. The results indicated that a maximum enrichment was obtained in the region of 10–20 kHz.

The research team which has advanced the field of DEP separation for cancer diagnosis is that of Becker, Gascoyne and co-workers (67), who used a flow cell and castellated electrode array (which they termed a "dielectric affinity column"), usually in conjunction with electrorotation for cell analysis. Interdigitated microelectrode arrays were used to retain human promyelocytic leukaemic cells (HL60) from blood collected from venipunctures in a ratio of 3:2. HL60 cells could subsequently be released by removal of the dielectrophoretic field and collected, while normal blood cells were eluted from the electrode chamber. This work on separation of cancer cells from blood was further explored by the same group (68) to demonstrate that the dielectric properties of metastatic human breast cancer cell line MDA-231 were significantly different from those of erythrocytes and T-lympho-cytes, and the differences were exploited to separate breast cancer cells from normal blood cells. Gascoyne and co-workers (15) continued the work on dielectrophoretic separation of cancer cells from blood. The "dielectric column" was used to separate and harvest viable HL60 and MDA-231 from whole blood. They concluded that cell surface morphology coupled with cell size dominated cell dielectrophoretic responses and acted as a sorting criterion. They have also recommended this technique for research areas where cell modification (such as that by stain or antibodies) may compromise the results or where specific markers or antibo-dies might not be available. Furthermore, since the parameters by which cells are identified in current sorting and characterisation techniques such as fluorescence-activated sorting, affinity column and centrifugal techniques play little or no role in AC-electrokinetics, they suggested the dielectrophoretic affinity column approach might be applied as an adjunct to the conventional methods to yield an overall improved discrimination.

The use of FFF for cancer studies has also been pursued by Gascoyne and co-workers. One example, used to separate cancer cells from stem cells, was performed by Huang and co-workers (1999) (69) by exploiting a combination of fluid flow and DEP. DEP field-flowfractionation (DEP-FFF) was used to purge human breast cancer MDA-435 cells from haemopoietic $CD34^+$ stem cells. A chamber with an array of interdigitated microelectrodes was used to generate DEP forces that levitated the cell mixture in a fluid flow profile. The $CD34^+$ cells were levitated higher and were carried faster by the fluid flow, earlier than the MDA-435. Using this set up as well as on-line flow cytometry, efficient separation of the cell mixture was observed in less than 12 min, with $CD34^+$ cell fractions being more than 99% pure. The authors suggested the potential usefulness of DEP-FFF in biomedical cell separation problems, including microfluid-scale-based diagnosis and provid-ing a preparative-scale purification of cell subpopulations.

Other groups are exploring new methods of cell separation for cancer applications. One novel approach was the use of cell levita-tion which was investigated later by a different group (70), where they described the use of an innovative printed circuit board (PCB) device to generate DEP. This was used to create software-controlled "cages" that can be moved to any place inside the microchamber and entrap cells according to their dielectrophore-tic properties. Different tumour cell lines were used, including Jurkat (T-lymphoid), erythroleukaemic (HEL) and melanoma (Colo38). The electrode array was suggested to form the basis for a lab on a chip, performing cell separation the basis of DEP levitation and movement of different tumour cells under different DEP conditions. Another example was the DEP-based Smart-Slide and the DEP array devices (71), studied for facilitating programmable interactions between microspheres and target tumour cells. The former device carried 193 parallel electrodes generating up to 50 cyliner shaped DEP cages, while the latter carried 102,400 arrayed electrodes that generated more than 10,000 spherical DEP cages. The results of the SmartSlide indicated that it can be used for transfection experiments in which target cells and microspheres are forced to share the same cage, thus leading to more efficient binding, while the array device allowed sequential and software controlled binding of individual and single microsphere to a single target tumour cell

6. Conclusion

Dielectrophoretic methods have been studied for over four decades, and extensively so from the 1990s. It has employed a wide range of applications and in different fields and hosts, such as plants, bacteria and viruses, as well as different mammalian systems, in particular those including samples of human origin. DEP has been used to move, trap, sort and separate, and analyse cells on the basis of characterising cells under different conditions, be it change of environment or drug treatment; the general aim has always involved obtaining a better understanding as to how cells or particles behave under certain conditions. Investigations like these have led the potential applications of DEP in areas like pharmacology, microbiology and oncology, some of which have made their way to commercial technology with many filed patents based on electrode design or a novel application. The most promising of all is the new merge of this technique, whether alone or in combination with others, into lab-on-a-chip systems.

References

1. Pohl, H. A. (1950) The motion and precipitation of suspenoids in divergent electric fields. *Journal of Applied Physics* 22, 869–871.

2. Pohl, H. A. (1978) *Dielectrophoresis*, Cambridge University Press, Cambridge.

3. Jones, T. B. (1995) *Electromechanics of Particles*, Cambridge University Press, Cambridge.

4. Zimmermann, U., Neil, G. A. (1996) *Electromanipulation of Cells*, CRC Press, Boca Raton.

5. Hughes, M. P. (2002) *Nanoelectromechanics in Engineering and Biology*, CRC press, Boca Raton.

6. Markx, G. H., Huang, Y., Zhou, X. F., Pethig, R. (1994) Dielectrophoretic characterisation of microorganisms. *Microbiology* 140, 585–591.

7. Mahaworasilpa, T. L., Coster, H. G. L., George, E. P. (1994) Forces on biological cells due to applied alternating (AC) electric fields. *Biochimica et Biophysica Acta* 1193, 118–126.

8. Huang, Y., Hölzel, R., Pethig, R., Wang, X. B. (1992) Differences in the AC electrodynamics of viable and non-viable yeast cells determined through combined dielectrophoresis and electrorotaion studies. *Physics in Medicine and Biology* 37, 1499–1517.

9. Falokun, C. D., Mavituna, F., Markx, G. H. (2003) AC electrokinetic characterisation and separation of cells with high and low embryogenic potential in suspension cultures of carrot (Daucus carota). *Plant Cell Tissue Organ Culture* **75**, 261–272.

10. Johari, J., Hübner, Y., Hull, J. C., Dale, J. W., Hughes, M. P. (2003) Dielectrophoretic assay of bacterial resistance to antibiotics. *Physics in Medicine and Biology* **48**, N193–N198.

11. Gimsa, J., Schnelle, T., Zechel, G., Glaser, R. (1994) Dielectric spectroscopy of human erythrocytes investigations under the influence of nystatin. *Biophysical Journal* **66**, 1244–1253.

12. Griffith, A. W., Cooper, J. M. (1998) Single-cell measurements of human neutrophil activation using electrorotation. *Analytical Chemistry* **70**, 2607–2612.

13. Markx, G. H., Talary, M. S., Pethig, R. (1994) Separation of viable and non viable yeast using dielectrophoresis. *Journal of Biotechnology* **32**, 29–37.

14. Stephens, M., Talary, M. S., Pethig, R., Burnett, A. K., Mills, K. I. (1996) The dielectrophoresis enrichment of CD34(+) cells from peripheral blood stem cell harvests. *Bone Marrow Transplantation* **18**, 777–782.

15. Gascoyne, P. R. C., Wang, X. B., Huang, Y., Becker, F. F. (1997) Dielectrophoretic separation of cancer cells from blood. *IEEE Transactions on Industry Applications* **33**, 670–678.

16. Yang, J., Huang, Y., Wang, X., Wang, X. B., Becker, F. F., Gascoyne, P. R. C. (1999) Dielectric properties of human leukocyte subpopulations determined by electrorotation as a cell separation criterion. *Biophysical Journal* **76**, 3307–3314.

17. Morgan, H., Hughes, M. P., Green, N. G. (1999) Separation of submicron bioparticles by dielectrophoresis. *Biophysical Journal* **77**, 516–525.

18. Borgatti, M., Altomare, L., Baruffa, M., Fabbri, E., Breveglieri, G., Feriotto, G., Manaresi, N., Medoro, G., Romani, A., Tartagni, M., Gambari, R., Guerrieri, R. (2005) Separation of white blood cells from erythrocytes on a dielectrophoresis (DEP) based 'Lab-on-a-chip' device. *International Journal of Molecular Medicine* **15**, 913–920.

19. Burt, J. P. H., Pethig, R., Gascoyne, P. R. C., Becker, F. F. (1990) Dielectrophoretic characterisation of friend murine erythroleukaemic cells as a function of induced differentiation. *Biochimica et Biophysica Acta* **1034**, 93–101.

20. Ratanachoo, K., Gascoyne, P. R. C., Ruchirawat, M. (2002) Detection of cellular responses to toxicants by dielectrophoresis. *Biochimica et Biophysica Acta* **1564**, 449–458.

21. Wang, X. J., Becker, F. F., Gascoyne, P. R. C. (2002) Membrane dielectric changes indicate induced apoptosis in HL-60 cells more sensitively than surface phosphatidylserine expression or DNA fragmentation. *Biochimica et Biophysica Acta* **1564**, 412–420.

22. Gascoyne, P., Mahidol, C., Ruchirawat, M., Satayavivad, J., Watcharasit, P., Becker, F. F. (2002) Microsample preparation by dielectrophoresis: isolation of malaria. *Lab on a Chip* **2**, 70–75.

23. Huang, Y., Wang, X. B., Becker, F. F., Gascoyne, P. R. C. (1996) Membrane changes associated with the temperature sensitive P85(gag-mos)-dependent transformation of rat kidney cells as determined by dielectrophoresis and electrorotation. *Biochimica et Biophysica Acta* **1282**, 76–84.

24. Hübner, Y., Hoettges, K. F., Hughes, M. P. (2003) Water quality test based on dielectrophoretic measurements of fresh water algae Selenastrum capricornutum. *Journal of Environmental Monitoring* **5**, 861–864.

25. Benguigui, L., Lin, I. J. (1982) More about the dielectrophoretic force. *Journal of Applied Physics* **53**, 1141–1143.

26. Pethig, R. (1991) Biological electrostatics: Dielectrophoresis and electrorotation. *Institute of Physics Conference Series* **118**, 13–26.

27. Broche, L. M., Labeed, F. H., Hughes, M. P. (2005) Extraction of dielectric properties of multiple populations from Dielectrophoretic collection spectrum. *Physics in Medicine and Biology* **50**, 2267–2274.

28. Pohl, H. A., Hawk, I. (1966) Separation of living and dead cells by dielectrophoresis. *Science* **152**, 647–649.

29. Gimsa J., Müller T., Schnelle, T., Fuhr, G. (1996) Dielectric spectroscopy of single human erythrocytes at physiological ionic strength: Dispersion of the cytoplasm. *Biophysical Journal* **71**, 495–506.

30. Pohl, H. A., Crane, J. A. (1971) Dielectrophoresis of cells. *Biophysical Journal* **11**, 711–727.

31. Tsoneva, I. C., Zhelev, D. V., Dimitrov, D. S. (1986) Red-blood-cell dielectrophoresis is axisymmetrical fields. *Cell Biophysics* **8**, 89–101.

32. Stoicheva, N. G., Dimitrov, D. S. (1986) Dielectrophoresis of myeloma cells, hybridomas and lymphocytes-B. *Dokladi na Bolgarskata Akademiya na Naukite* **39**, 105–107.

33. Stoicheva, N., Dimitrov, D. S. (1986) Dielectrophoresis of single protoplasts, chloroplasts and lymphocytes in axisymmetrical fields. *Biophysical Journal* **49**, A92–A92.

34. Arnold, W. M., Zimmermann, U. (1982) Rotation of an isolated cell in a rotating field. *Naturwissenschaften* **69**, 297–298.

35. Arnold, W. M., Zimmermann, U. (1982) Rotating field induced rotation and measurement of the membrane capacitance of single mesophyll cells of *Avena-Sativa*. *Zeitschrift fur Naturforschung-c- A Journal of Biosciences* **37**, 908–915.

36. Arnold, W. M., Zimmermann, U. (1988) Electrorotation: development of a technique for dielectric measurements on individual cells and particles. *Journal of Electrostatics* **21**, 151–191.

37. Hanai, T. (1960) Theory of the dielectric dispersion due to the interfacial polarisation and its application to emulsion. *Kolloid Z* **171**, 23–31.

38. Irimajiri, A., Hanai, T., Inouye, V. A. (1979) Dielectric theory of "multi-stratified shell" model with its application to lymphoma cell. *Journal of Theoretical Biology* **78**, 251–269.

39. Chan, K. L., Gascoyne, P. R. C., Becker, F. F., Pething, R. (1997) Electrorotation of liposomes: verification of dielectric multi-shell model for cells. *Biochimica et Biophysica Acta* **1349**, 182–196.

40. Pethig, R., Huang, Y., Wang, X.-B., Burt, J. P. H. (1992) Positive and negative dielectrophoretic collection of colloidal particles using interdigitated castellated microelectrodes. *Journal of Physics D: Applied Physics* **24**, 881–888.

41. Gascoyne, P. R. C., Huang, Y., Pethig, R., Vykoukal, J., Becker, F. F. (1992) Dielectrophoretic separation of mammalian cells studied by computerised image analysis. *Measurement Science and Technology* **3**, 439–445.

42. Zhou, X. F., Markx, G. H., Pethig, R., Eastwood, I. M. (1995) Differentiation of viable and nonviable bacterial biofilms using electrorotation. *Biochimica et Biophysica Acta* **1245**, 85–93.

43. Quinn, C. M., Archer, G. P., Betts, W. B., O'Neill, J. G. (1996) Dose dependent dielectrophoretic response of *Cryptosporidium* oocysts treated with ozone. *Letters in Applied Microbiology* **22**, 224–228.

44. Betts, W. B., Brown, A. P. (1999) Dielectrophoretic analysis of microbes in water. *Journal of Applied Microbiology* **85**, 201S–213S.

45. Dalton, C., Goater, A. D., Drysdale, J., Pething, R. (2001) Parasite viability by electrorotation. *Colloids and Surfaces A-Physicochemical and Engineering Aspects* **195**, 263–268.

46. Fatoyinbo, H. O., Hughes, M. P., Martin, S. P., Pashby, P., Labeed, F. H. (2007) Dielectrophoretic separation of Bacillus subtilis spores from environmental diesel particles. *Journal of Environmental Monitoring* **9**, 87–90.

47. Suheiro, J., Yatsunami, R., Hamada, R., Hara, M. (1999) Quantitative estimation of biological cell concentration suspended in aqueous medium by using dielectrophoretic impedance measurement method. *Journal of Physics D: Applied Physics* **32**, 2814–2820.

48. Suheiro, J., Shutou, M., Hatano, T., Hara, M. (2003) High sensitive detection of biological cells using dielectrophoretic impedance measurement method combined with electropermeabilisation. *Sensors and Actuators B-Chemical* **96**, 144–151.

49. Ali, M. A., Bashier, S. A. (2006) Effect of fast green dye on some biophysical properties of thymocytes and splenocytes of albino mice. *Food additives and contaminants* **23**, 452–461.

50. Minerick, A. R., Zhou, R. H., Takhistov, P., Chang, H.-C. (2003) Manipulation and characterisation of red blood cells with alternating current fields in microdevices. *Electrophoresis* **24**, 3703–3717.

51. Ermolina, I., Polevaya, Y., Feldman, Y. (2000) Analysis of dielectric spectra of eukaryotic cells by computer modelling. *European Biophysics Journal with Biophysics Letters* **29**, 141–145.

52. Neu, B., Georgieva, R., Meiselman, H. J., Baumler, H. (2002) Alpha- and beta-dispersion of fixed platelets: comparison with a structure-based theoretical approach. *Colloids and Surfaces A-Physicochemical and engineering aspects* **197**, 27–35.

53. Chan, K. L., Morgan, H., Morgan, E., Cameron, I. T., Thomas, M. R. (2000) Measurements of the dielectric properties of peripheral blood mononuclear cells and trophoblast cells using AC-electrokinetic techniques. *Biochimica et Biophysica Acta* **1500**, 313–322.

54. Hu, X., Arnold, W. M., Zimmermann, U. (1990) Alterations in the electrical properties of lymphocyte T and lymphocyte B membranes induced by mitogenic stimulation-activation monitored by electrorotation of single cells. *Biochimica et Biophysica Acta* **1021**, 191–200.

55. Gambari, R., Borgatti, M., Altomere, L., Manaresi, N., Medoro, G., Romani, A., Tartagni, M., Guerrieri, R. L. (2003) Applications to cancer research of "lab-on-a-chip" devices based on dielectrophoresis (DEP). *Technology in Cancer Research and Treatment* **2**, 31–40.

56. Mischel, M., Rouge, F., Lamprecht, I., Lamprecht, I., Aubert, C., Prota, G. (1983) Dielectrophoresis of malignant human melanocytes. *Archives of Dermatological Research* **275**, 141–143.

57. Cristofanilli, M., De Gasperis, G., Zhang, L. S., Hung, M.-C., Gascoyne, P. R. C., Hortobagyi, G. N. (2002) Automated electrorotation to reveal dielectric variations related to HER-2/neu overexpression in MCF-7 sublines. *Clinical Cancer Research* **8**, 615–619.

58. Wang, X.-B., Huang, Y., Gascoyne, P. R. C., Becker, F. F., Hölzel, R., Pethig, R. (1994) Changes in friend murine erythroleukaemia cell membranes during induced differentiation determined by electrorotation. *Biochimica et Biophysica Acta* **1193**, 330–344.

59. Labeed, F. H., Coley, H. M., Thomas, H., Hughes. M. P. (2003) Assessment of multidrug resistance reversal using dielectrophoresis and flow cytometry. *Biophysical Journal* **85**, 2028–2034.

60. Coley, H. M., Labeed, F. H., Thomas, H., Hughes, M. P. (2007) Biophysical characterization of MDR breast cancer cell lines reveals the cytoplasm is critical in determining drug sensitivity. *Biochimica et Biophysica Acta* **1770**, 601–608.

61. Labeed, F. H., Coley, H. M., Hughes, M. P. (2006) Differences in the biophysical properties of membrane and cytoplasm of apoptotic cells revealed using dielectrophoresis. *Biochimica et Biophysica Acta* **1760**, 922–929.

62. Chin, S., Hughes, M. P., Coley, H. M., Labeed, F. H. (2006) Dielectrophoresis is a rapid technique for detecting early apoptosis. *International Journal of Nanomedicine* **1**, 333–337.

63. Huang, Y., Wang, X.-B., Hölzel, R., Becker, F. F., Gascoyne, P. R. C. (1995) Electrorotational studies of the cytoplasmic dielectric properties of Friend murine erythroleukaemia cells. *Physics in Medicine and Biology* **40**, 1789–1806.

64. Wang, X. J., Yang, J., Gascoyne, P. R. C. (1999) Role of peroxide in AC electrical field exposure effects on Friend murine erythroleukemia cells during dielectrophoretic manipulations. *Biochimica et Biophysica* **1426**, 53–68.

65. Markx, G. H., Rousselet, J., Pethig, R. (1997) DEP-FFF: Field-flow fractionation using non-uniform electric fields. *Journal of Liquid chromatography & Related Technologies* **20**, 2857–2872.

66. Talary, M. S., Mills, K. I., Hoy, T., Burnett, A. K., Pethig, R. (1995) Dielectrophoretic separation and enrichment of CD34+ cell subpopulation from bone marrow and peripheral blood stem cells. *Medical & Biological Engineering & Computing* **33**, 235–237.

67. Becker, F. F., Wang, X. B., Huang, Y., Pethig, R., Viykoukal, J., Gascoyne, P. R. C. (1994) The removal of human leukaemia cells from blood using interdigitated microelectrodes. *Journal of Applied Physics D* **27**, 2659–2662.

68. Becker, F. F., Wang, X.-B., Huang, Y., Pething, R., Vykoukal, J. (1995) Separation of human breast cancer cells from blood by differential dielectric affinity. *Proceedings of the National Academy of Sciences of the United States of America* **92**, 860–864.

69. Huang, Y., Yang, J., Wang, X.-B., Becker, F. F., Gascoyne, P. R. C. (1999) The removal of human breast cancer cells from haemopoetic CD34+ stem cells by dielectrophoretic field flow fractionation. *Journal of Hematotherapy & Stem Cell Research* **8**, 481–490.

70. Altomare, L., Borgatti, M., Medoro, G., Manaresi, N., Tartagni, M., Guerrieri, R., Gambari, R. (2003) Levitation and movement of human tumour cells using a printed circuit board device based on software controlled dielectrophoresis. *Biotechnology and Bioengineering* **82**, 474–479.

71. Borgatti, M., Altomare, L., Abonnec, M., Fabbri, E., Manaresi, N., Medoro, G., Romani, A., Tartagni, M., Nastruzzi, C., Di Croce, S., Tosi, A., Mancini, I., Guerrieri, R., Gambari, R. (2005) Dielectrophoresis-based 'Lab-on-a-chip' devices for programmable binding of microspheres to target cells. *International Journal of Oncology* **27**, 1559–1566.

Chapter 10

Wireless Endoscopy: Technology and Design

David R. S. Cumming, Paul A. Hammond, and Lei Wang

Abstract

In this chapter we review the current capsule technology and the more conventional "gold standard" technologies against which the wireless devices are compared. Over the years there have been several implementations of capsule devices of growing sophistication as new technology has become available. A notable feature is the extent to which the devices available at any given time have relied upon other more mainstream technologies from which capsule builders have been able to borrow. As an inevitable consequence, device complexity and functionality have increased.

Key words: Lab on a pill, wireless, diagnostics, capsule, sensor.

1. Introduction

"It is rather inconvenient to swallow a physician." So begins one of the first papers on radiotelemetry capsules by Mackay (1). However, to diagnose a wide range of conditions, it is extremely useful to be able to make measurements of the conditions inside a patient's gut. Conventional methods of measuring inside the gastrointestinal (GI) tract require the passage of tubes through the mouth, nose or anus. Not only are these unpleasant procedures for the patient, but the physical presence of the tube can affect the phenomenon being studied. Furthermore, the far reaches of the GI tract such as the distal small intestine and proximal colon are almost inaccessible from either orifice. A device that could be swallowed and would transmit the required measurements without wires, from inside the patient, is clearly desirable.

In this chapter we review the current capsule technology and the more conventional "gold standard" technologies against which the wireless devices are compared. **Table 10.1** is a summary

M.P. Hughes, K.F. Hoettges (eds.), *Microengineering in Biotechnology,* Methods in Molecular Biology 583,
DOI 10.1007/978-1-60327-106-6_10, © Humana Press, a part of Springer Science+Business Media, LLC 2010

Table 10.1
A comparative summary of wireless devices

Device	Descriptions	Objective	RF (MHz)	Size (mm)	Comments
SMART pill	Intra-luminal pH measurement, tubeless gastric analysis, an alternative for pentagastrin tests.	Pressure, pH	1.98	20.0 × 8.0	Upgraded Heidelberg-type pH pill*
M2A	Imaging within GI tract, positive for the inspection of IBDs and Crohn's disease.	Image	434	26.0 × 11.0	GivenImaging (2)
NASA pill	Core-body temperature measurement, identify temperature changes within a potentially ulcerated area.	Temperature, pressure	174–216	30.0 × 10.0	Also Implantable (3)
BRAVO	Catheter-free oesophageal pH monitoring, GERD diagnosis	pH	Unknown	26.0 × 6.3	Medtronic (24)
InteliSite	Assessing drugs absorption within a specific region of the GI tract	N/A	6.78	35.0 × 10.0	Clinical evaluation stage (4, 5)
GI micro-robot	Bi-directional wireless endoscopes with real-time imaging and process	Image	434	40.0 × 11.0	Intelligent Micro-system Centre, Korea (34)
MMM	Magnetic markers as a non-invasive tool to monitor GI transit	Location	N/A	26.0 × 6.3	Freie university Berlin, Germany (6)

(continued)

Table 10.1 (continued)

Device	Descriptions	Objective	RF (MHz)	Size (mm)	Comments
Inchworm	Semi-mechanic device can be externally controlled for in-gut localisation	N/A	N/A	30.0 × 10.0	University of Pisa, Italy (7)
Medical actuator	Capsule-shaped magnetic actuator and capsule micro pump can be used inside GI tract	N/A	N/A	40.0 × 11.0 (magnetic) 23.0 × 13.0 (micropump)	Japanese research (8, 9)
IDEAS	Bringing together System-on-Chip and Lab-on-a-Chip methodologies to develop a GI monitoring microsystem.	Temperature, conductivity, pH, dissolved oxygen	30	Various	UK research (40)
Drug pill	The largest conventional capsule for oral drugs delivery	N/A		23.3 × 8.2	Size 00 pill

*The Heidelberg device is one of the earliest demonstrations of a pH measuring capsule (10).

of some of the better known devices. Over the years there have been several implementations of capsule devices of growing sophistication as new technology has become available. A notable feature is the extent to which the devices available at any given time have relied upon other more mainstream technologies from which capsule builders have been able to borrow. As an inevitable consequence, device complexity and functionality have increased.

The design of a wireless endoscopic capsule is a multidisciplinary task, requiring skills in sensors design (physical and chemical), electronics, wireless, laboratory-on-a-chip, portable power.

2. Traditional Diagnostic Techniques

The traditional examination of the GI tract is performed by both invasive and non-invasive procedures.

2.1. Non-invasive Techniques

The terminology "non-invasive" relates to a technique that does not involve puncturing the skin or entering a body cavity, such as radiography (X-ray), computed axial tomography (CAT), positron emission tomography (PET), ultrasonography and nuclear magnetic resonance imaging (MRI) (11). Non-invasive techniques such as radiography and ultrasonography are useful tools in the investigation of GI disease conditions if one wants to locate foreign bodies, intussusceptions and abnormal masses such as tumours. There is the added advantage that no sedation is required for the procedure, although the use in veterinary medicine requires some form of sedation (often a general anaesthetic). However, X-ray and ultrasonography suffers from poor image quality of soft tissues and are obstructed by the presence of solid ingesta, faecal material or, in the case of ultrasonography, the accumulation of intraluminal gas. CAT and MRI scans offer enhanced images of soft tissues through the manipulation of X-rays (in CAT) or utilizing the magnetic orientation of atoms by pulsating electromagnetic radiation. PET scans offer some degree of functional assessment of internal organs/structures through the observation of physiological processes in living tissues. However, these methods are not able to provide any information about the internal biochemical environment of the gut.

2.2. Invasive Techniques

Since the above non-invasive methods cannot present any direct information on the biochemical environment within the GI tract, endoscopy, a relatively invasive technology, has gained clinical acceptance to view the GI tract and perform biopsies for later analysis. Currently gastroscopy and colonoscopy are used as the main tool to observe suspected areas of inflammation and to detect the presence of foreign objects or tumours (12). However, endoscopy is a time consuming procedure, is potentially hazardous, can cause damage to the intestinal wall and, in particular, it is difficult to access the distal small intestine. Sedation or a general anaesthetic is often required for animal endoscopy, and the considerable inconvenience and irritation of this technique discourages human patients from undergoing the procedure. The technique is also unsuitable for monitoring GI dysfunction since it cannot measure the GI tract in real-time over an extended period (13).

3. Applications for Wireless Capsule Devices

3.1. Human Medicine

Medical devices must first satisfy the requirement for clinical efficacy. Because of this the capsule devices that have been developed to date have only a simple range of sensor capabilities due to the relative lack of medical data that would normally encourage the development of more sophisticated tools. In some regards the small amount of data available is due to the historical lack of suitable tools, especially with respect to the distal small intestine, although increased acceptance of capsule technology is gradually addressing this problem. At present the dominant measurement capabilities are image acquisition, temperature and pH, although devices capable of sample retrieval, pressure sensing and dissolved oxygen measurement, amongst others, have been explored.

In 1957, the first two radiotelemetry capsules were developed independently in Stockholm (14) and New York (15). They measured approximately 10 mm in diameter and 30 mm in length.

Both were designed to measure pressure using a diaphragm to move an iron core inside a coil. The coil was the tuning element of a single-transistor oscillator circuit, the frequency of which therefore, depended on the pressure. The circuit used by Mackay and Jacobson is shown in **Fig. 10.1** (14). It produced bursts of oscillations in the 100 kHz range; the frequency was a function of pressure, and the burst repetition rate a function of temperature. This ingenious circuit was capable of measuring both pressure and temperature, which were recorded using a standard radio receiver.

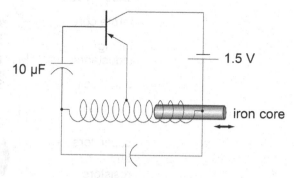

Fig. 10.1. Circuit diagram of a pressure-sensitive single-transistor oscillator.

pH is one of the most important parameters to measure in any biological system. In terms of pH measurement, capsule-based systems have been employed most successfully to diagnose gastro-oesophageal reflux disease (GERD) (16). These systems require a tethered capsule to monitor whether stomach acid is refluxing back into the oesophagus and causing the burning sensation commonly known as heartburn. Measurements of pH in the GI tract using "flow-through" capsules have also been used to study inflammatory bowel diseases (IBDs) such as Crohn's disease and ulcerative colitis (17, 18). Knowledge of the pH profile along the GI tract has also led to the development of drug capsules which dissolve in alkaline environments, ensuring that the drug is only released beyond the acid environment of the stomach.

In 1959, Wolff (19) began work to develop radiotelemetry capsules that could be mass produced in order to reduce the cost and increase the number available for clinical trials. As well as a pressure-sensitive capsule, his laboratory developed capsules for measuring temperature and pH, all based on single-transistor oscillators. The temperature-measuring capsule used a coil wound on a core made from a nickel-iron alloy, which exhibited a large change in permeability with temperature. This changed the inductance of the tuning coil and shifted the frequency of transmission in a manner similar to the pressure sensor of Mackay (14). The pH-measuring capsule used a glass electrode connected to a silicon diode, which operated as the tuning element for the oscillator. The potential of the electrode varied with pH and this voltage changed the capacitance of the diode. A reference electrode, needed to complete the pH circuit, was also included in the capsule which is shown in **Fig. 10.2**.

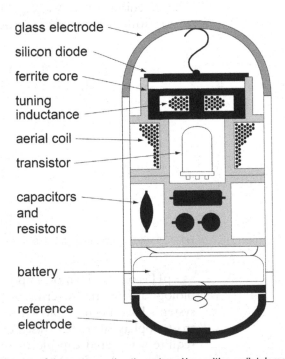

Fig. 10.2. Diagram of the cross-section through a pH-sensitive radiotelemetry capsule.

A similar pH-sensitive capsule based on a glass electrode was described by Watson and Kay (20). It was used in the first medical study (of 9 subjects) to plot the pH profile along the entire length of the GI tract (21). Localisation of the capsule was performed by taking four or five X-rays during the study. The results from the capsule were compared with readings from faecal samples using a standard pH electrode. In all cases, the discrepancy was found to be less than 0.2 pH units.

These early capsules were prone to failure due to the ingress of moisture through the epoxy, which was used to seal the glass electrode in place. An improved design that housed the battery and electronics inside the glass electrode was developed by Colson (22). The receiving antenna array was embedded in a cloth band worn around the waist. Data was collected on a portable solid-state recorder that allowed the patient to carry out their normal activities. This equipment was used in a much larger study of 66 subjects (23). During this study, the capsule was located (by the subject) in one of nine possible segments, using a highly directional aerial attached to a signal-strength meter. Coupled with obvious changes in pH, such as the transition from stomach (strong acid) to the small intestine (mild alkaline), this allowed an accurate plot to be made of the pH profile along the GI tract (**Fig. 10.3**).

For conditions such as GERD, it is necessary to measure the pH at a fixed location, rather than measuring a flow-through profile. Early studies achieved this by "lowering" the capsule into

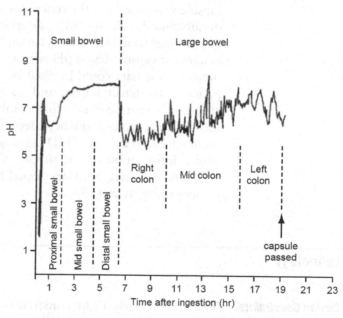

Fig. 10.3. Graph of a typical gastrointestinal pH profile obtained by radiotelemetry. Reproduced from (23) with permission from the BMJ Publishing Group.

position on a piece of thread, and fixing the free end to the subject's cheek when the capsule was in position. More recently, a capsule has been developed that is temporarily anchored to the wall of the GI tract using an endoscopic delivery system (24) that is relatively invasive. A vacuum pump sucks tissue into a well in the capsule and a pin is pushed through the tissue holding it in place after the delivery system has been removed. This capsule is 6 mm in diameter and 25 mm long, and it uses an antimony pH electrode. It transmits to a pager-sized receiver, allowing patients to continue their normal activities without restriction of diet or exercise.

Commercial radiotelemetry capsules have been developed that have the potential to replace conventional fibre-optic endoscopy and colonoscopy. The M2A capsule from Given Imaging Ltd. (2) contains a single-chip complementary metal-oxide-semiconductor (CMOS) image sensor, an application-specific integrated circuit (ASIC) for video transmission, and white light-emitting diodes (LEDs) for illumination. The data is transmitted to an array of eight antennae worn on a belt that also allow the capsule's position to be localised, it is claimed, to within 3.8 cm. The system is particularly well suited to detection of bleeding and the software contains blood-recognition algorithms to automatically highlight suspect areas. To date it has been used in numerous clinical trials and as a diagnostic tool in about 4,000 patients (2, 25).

3.2. Animal Applications

Data-logging telemetry capsules have been used to measure the pH inside the stomach small animals (26), including penguins (27). The same technology has been used in cattle where the capsule was located in the reticulum to measure the effect of diet on subclinical acidosis (28). (The reticulum is the second stomach of a ruminant.) In fact, livestock monitoring may well be the major market for capsule-based pH sensors. When combined with temperature, the data could be used by farmers to optimise feeding patterns, to detect illness, and to manage breeding. There is already a system available that combines temperature measurement with a radio-frequency identification (RFID) chip that is unique to each animal. RFID tagging of livestock provides guaranteed information on the supply of meat from "farm-to-fork," and standards have been developed for the manufacture of such devices and systems (29).

4. Technology

4.1. Design Constraints

There are clearly tight constraints placed on the design and implementation of a capsule-based diagnostic system. The overall dimensions should be small enough to allow the device to pass through all the GI sphincters with relative ease, including the lower oesophageal sphincter and the pyloric sphincter. The capsule must also be cheap to make since it will only be used once. Low power consumption is a requirement to minimise battery, hence overall, size and increase operating time. Gastric emptying takes between 30 min and 4 h to complete and semi-digested food (chyme) takes 2–3 h to pass through the small intestine. Once in the colon, gut content moves relatively slowly at approximately 5–10 cm/h. Peristaltic waves in the colon are known as mass movements and only occur 1–3 times per day. Overall the capsule might take a maximum of 8 h to traverse the upper alimentary tract and the small intestine, while a complete passage through the GI tract might take up to 32 h. Using readily available silver oxide battery technology, an energy storage density of 500 mWh/mL can be achieved (30), thus a suitable source, such as two SR48 cells (110 mWh each) could deliver enough energy to complete small intestinal measurements if the power consumption was less than 20 mW.

The data sampled in the GI tract by a capsule must be retrieved accurately and securely. This usually means that the data must be wirelessly transmitted and correctly received by a device worn by the patient. There are a number of radio communication standards encompassing several international industrial, scientific and medical telemetry bands (pan-European medical device frequency allocations (31), and the US Federal Communications Commission frequency allocations for biomedical telemetry and industrial, scientific and medical (ISM) devices – regulations S5.150, US209 and US350). The main bands of interest are at 418 MHz, 434 MHz, 868 MHz and 915 MHz.

As with all measuring devices, the user must be confident that the data retrieved is accurate. The problems of accuracy can be dealt with via the normal techniques of instrument design and calibration. However, an additional constraint for wireless devices is that the data must be secure. Since capsule devices operate in the unlicensed ISM frequency bands there is a severe risk of interference that could be particularly dangerous in the context of a medical device. Of necessity, the sensors and signal acquisition electronics require analogue circuits. In the early devices the entire design was analogue, making the data transfer from the devices extremely insecure. However, modern electronic techniques permit designers to convert the analogue signal to the digital domain within the capsule, enabling the use of secure digital wireless techniques. These techniques ensure that data from any given capsule can be uniquely identified to avoid attributing diagnostic information to the wrong individual. Details of such designs, and others concerning wireless sensor systems, may be found in the literature (32, 33).

4.2. Capsule System Design

Whilst there is no single way to build a capsule system using current technology, many of the aspects of a complete solution are illustrated in **Fig. 10.4**. In the capsule, the sensors are the data gathering devices that are connected to the electronic instrumentation required to acquire the signal. In the earlier devices all the data was managed by analogue electronics, but more recent devices convert the signals into a digital representation. In this way a common platform can be developed in which one basic controller design can be reused for successive products or different sensor modalities. This is an example of the system-on-chip (SoC) methodology, in which the majority of the components are connected together on a single chip. Commonplace examples of products containing SoC devices are mobile telephones, digital television and radio receivers, and computer game consoles. The design of small SoCs is well suited to capsule design since the device requirements are both complex and unusual to the extent that a small enough device can not be easily assembled from off-the-shelf components.

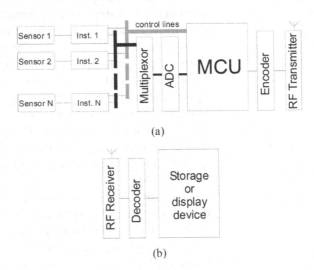

Fig. 10.4. A typical block diagram of the key features of (**a**) a capsule (**b**) a receiver. Because the receiver is not constrained by power and size it can have any reasonable level of complexity.

For capsule devices, the use of a digital architecture enables substantially more complex systems to be built that are capable of performing many measurements. **Figure 10.5** shows an integrated circuit designed for use in a micro-capsule system that contains all the electronics required. In this implementation, a small low power transmitter has been integrated on to the same chip with a usable operating range of only 10–20 cm. Because of the difficulty of building a suitable transmitter onto the same integrated circuit as the rest of the instrument, it is usual to have

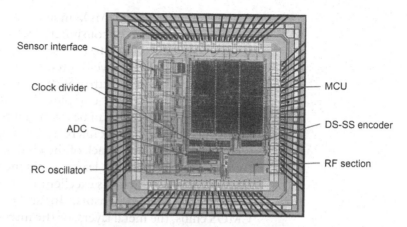

Sensor interface

Clock divider

ADC

RC oscillator

MCU

DS-SS encoder

RF section

Fig. 10.5. A photo-micrograph of an integrated circuit providing nearly all the electronics for a capsule device.

a separate RF section, usually made from commercially available parts. The majority of devices have only a one-way wireless link to enable data transmission to an external device. However, a device providing a two-way link has recently been demonstrated (34). The advantages of a two-way wireless link are improved security of the wireless connection and external control of the capsule.

Once all the measurements are combined, they are encoded and transmitted over a wireless link to a receiver outside the body. There are many possible ways of building the external system and for illustration purposes we show a system combining an RF section, a decoder and display or data storage unit. This latter unit, usually a wearable device, could simply record data on to storage media for subsequent analysis, or provide real-time display and analysis capabilities. Another possibility that has been investigated is implementing the external unit as a web-server enabling clinicians to "look-in" from potentially anywhere with a web-browser (35).

4.3. Integrated Sensors

There are a wide variety of microsensor technologies now available, many of which can be applied to diagnostic capsule applications. A useful text describing many examples is given by Gardner (36). Microsensors, and in particular, those sensors that can be integrated in to a small format sharing a common platform with the electronics, are very useful for size-constrained systems such as a diagnostic capsule.

4.3.1. Physical

One of the most significant examples has been the use of CMOS video chips by Given Imaging (2). In addition to the ability to integrate the electronics and the sensor on the same chip, the advantages of the CMOS video approach, as opposed to using a charge coupled device, are the relatively low cost and the ability of the device to operate at relatively low voltage.

The integration of CMOS image sensors with electronics is a result of the advance of consumer electronics. Other integrated sensors, targeted at industrial and medical applications, have also been developed that are well suited to capsule systems (37). With the advent of micro electro-mechanical systems (MEMS), it is now possible to pattern complex 3-D structures into CMOS chips. MEMS processes can be divided into either bulk or surface micro-machining. In bulk micro-machining, the silicon substrate is etched away from the back of the chip or wafer, using the oxide layer as an etch stop. This leaves a thin membrane containing the CMOS circuits, which has excellent thermal isolation and can be used for heat-based sensors. In surface micro-machining of CMOS chips, the metal layers, or the inter-metal dielectric layers, are etched away to leave free-standing structures. Resonating beams for mass sensing or thin filaments for heat sensing can be made in this way. Several sensors have been fabricated using a combination of CMOS and MEMS technologies. For example, a recent capsule-type device for measuring intra-vascular pressure uses a MEMS capacitive pressure sensor integrated onto a CMOS chip (38, 39).

4.3.2. Chemical

As already discussed, pH sensors can be made using conventional glass electrode methods, but the arrival of chip-based sensors has enabled a more integrated approach to be adopted. Using this method it has been possible to implement more than one sensor on a single chip hence increasing functionality whilst contributing to the overall aim of reducing the capsule size. **Figure 10.6** shows two sensor chips that have been developed for a laboratory-in-a-pill device (LIAP) (40). The chips contain a diverse range of sensor technology, not least of which is a microfabricated (Ag/AgCl) reference electrode.

Chemical sensors have also been realised on CMOS chips. They may be classified as follows (37):

chemo-mechanical sensors typically use a polymer-coated resonating beam whose fundamental frequency is changed by the mass of absorbed gas molecules;

catalytic sensors have an electrically heated suspended filament that causes local oxidation reactions, and measures the heat loss as a change in temperature;

thermoelectric sensors use thermocouples to measure heat liberated or consumed by the interaction of a membrane with an analyte;

optical sensors use photodiodes to measure the light output from bioluminescent bacteria when they metabolise the target compound;

voltammetric sensors are miniaturised versions of the standard 3-electrode cell to measure the electron exchange currents that occur in redox reactions;

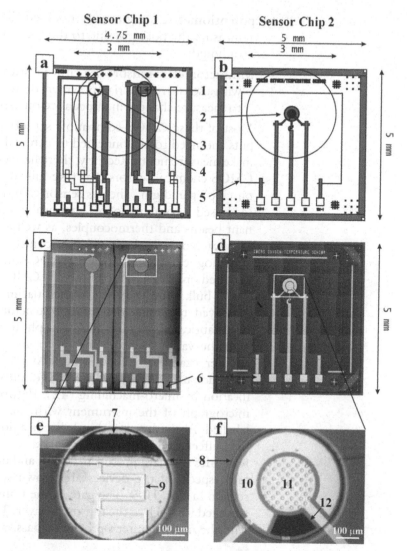

Fig. 10.6. Two sensor chips developed for a laboratory-in-a-pill. (**a**) schematic diagram of *Chip 1*, measuring 4.75×5 mm^2, comprising a pH based on an ion-sensitive field effect transistor (ISFET) sensor (1), a dual electrode conductivity sensor (3) and a silicon diode temperature sensor (4); (**b**) schematic diagram of *Chip 2*, measuring 5×5 mm^2, comprising an electrochemical oxygen sensor (2) and a Pt resistance thermometer (5). Once integrated in the pill, the area exposed to the external environment is illustrated by the 3 mm diameter circle; (**c**) photomicrograph of sensor *Chip 1* and (**d**) sensor *Chip 2*. The bonding pads (6), which provide electrical contact to the external electronic control circuit, are shown; (**e**) close up of the pH sensor consisting of the integrated 3×10^{-2} mm^2 AglAgCl reference electrode (7), a 500 μm diameter and 50 μm deep, 10 nL, electrolyte chamber (8) defined in polyimide, and the 15×600 μm floating gate (9) of the ISFET sensor; (**f**) an oxygen sensor is likewise embedded in an electrolyte chamber (8). The three-electrode electrochemical cell comprises the 1×10^{-1} mm^2 counter electrode (10), a micro-electrode array of 57×10 μm diameter (4.5×10^{-3} mm^2) working electrodes (11) defined in 500 nm thick PECVD Si$_3$N$_4$, and an integrated 1.5×10^{-2} mm^2 AglAgCl reference electrode (12). Reproduced from (40), © 2004 IEEE.

potentiometric sensors use modified field-effect transistors to measure the potential due to the concentration of ions in a gas or liquid;

conductometric sensors use either resistors or capacitors coated with a sensitive material (polymers or metal oxides, for example) to measure changes in impedance on exposure to the analyte.

Most of these CMOS-compatible sensors produce analogue outputs and need to be connected to external equipment in order to make measurements. Recently, there have been several examples of CMOS chemical sensors that take full advantage of the "system-on-chip" paradigm. The gas sensor chip described by Hagleitner (41) used a combination of chemically sensitive capacitors, resonant beams and thermocouples, as well as a temperature sensor, integrated on a single chip. All the control and sensing electronics, an analogue to digital converter (ADC) and a digital interface were included on the chip. A commercial CMOS process was used, and both bulk and surface micro-machining techniques were employed to define the sensing structures, after the chip had been fabricated. **Figure 10.7** is a photomicrograph of the chip with the various sensors and electronic circuits highlighted. In another example, a fully integrated pH measuring instrument was made using a standard CMOS foundry process with no modification by micro-machining (42). **Figure 10.8** shows a photomicrograph of the instrument with the individual components labelled. At the heart of this device is a floating gate ion-sensitive field effect transistor (ISFET) made by taking advantage of the foundries standard process materials and design rules. The important aspects of the devices are shown schematically as a cross-section in **Fig. 10.9**. The gate of the transistor is connected by a so-called via stack to the top-metal layer. The top-metal is covered with the manufacturer's protective passivation layer that is silicon

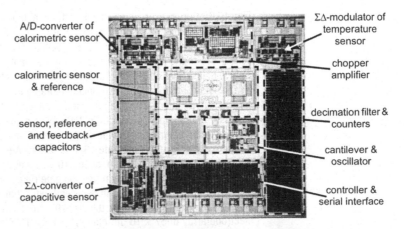

Fig. 10.7. Micrograph of system-on-chip CMOS gas sensor with the sensing components and the interface electronics highlighted. Reproduced from (41), © 2002 IEEE.

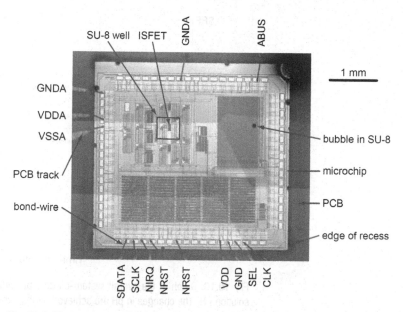

Fig. 10.8. Photomicrograph of encapsulated system-on-chip pH meter.

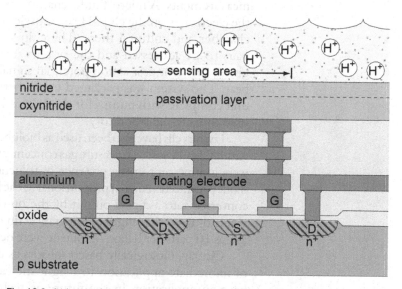

Fig. 10.9. A sketch of the cross-section through a floating gate CMOS ISFET.

nitride, a well-known pH-sensing membrane. **Figure 10.10** is a graph of the digital output from the chip (left-hand scale) calibrated to pH (right-hand scale).

4.3.3. Biological

Another example, this time of a "partial SoC" CMOS chemical sensor, uses living cells as the transducer to detect the presence of a toxin (43). It is not a complete system-on-chip as it requires off-chip analogue electronics and a micro-controller to make

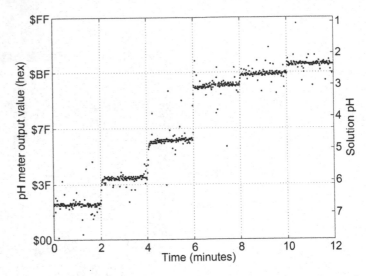

Fig. 10.10. Graph of the overall system-on-chip pH meter response to changes in the solution pH. The changes in pH are achieved by the addition of HCl to the test solution.

measurements. A micro-fluidic chamber is clamped in place above the electrodes on the chip. Heart muscle cells are injected into the chamber and cultured there. The chip allows different electrode pairs to be addressed and the system automatically selects those giving the strongest action potential signals from the cells as they beat. The system was packaged into a battery-powered handheld unit complete with pumps for the micro-fluidics, allowing it to be used outside the laboratory.

Living cells have also been used as bioluminescent "bio-reporters" with a CMOS SoC, to measure gas concentrations (44). The cells used were luminescent in the presence of toluene. An integrated photodiode produced a current proportional to the light intensity, which was converted into a digital output by the on-chip processing circuitry. Depending on the length of integration time used, concentrations as low as 10 parts per billion of toluene were detected.

Clearly, biologically based sensors as described above are not of immediate application to diagnostic capsule devices, but may have an application in the future as technology moves towards highly specific discriminatory techniques.

5. System Design Methodology

In addition to having all the required components for the implementation of a capsule device, one must think about how the complete system will be designed to achieve the desired performance. As we have seen there is rapid progress away from bench-top instrumentation design to modern integrated circuit implementations. As a consequence it is appropriate to use silicon design methodologies. Such methodologies have emerged from the microelectronics industry as more complex designs have been required (45, 46). We present a methodology here that is relatively simple by the standards of the industry, but encapsulates sufficient detail to enable accurate design of a sensor system.

5.1. Analogue Electronic Front-End Signal Acquisition Design

The steps involved in designing an analogue circuit are illustrated in **Fig. 10.11**. The main difference from digital design is that both the schematic and physical designs are created by hand. An analogue circuit starts off as a high-level model or simply a list of requirements that must be met. The first attempt at a circuit design is made using a schematic editor to draw a diagram of the components and their connections. The standard components available in a CMOS process are MOSFETs, resistors and capacitors, but others such as diodes, bipolar transistors and inductors can also be created. Parameterised models for all available component types, obtained by characterising fabricated devices, are provided by the foundry. The circuit is simulated by an analogue circuit simulator such as SPICE.

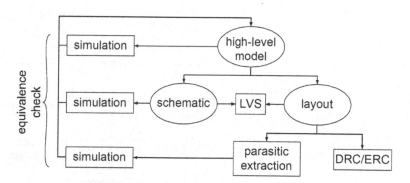

Fig. 10.11. Flow-chart for the computer-aided analogue circuit design process.

The circuit is unlikely to fulfil its requirements at the first attempt, so either the topology of the circuit or the parameters of its components are changed. Depending on the complexity of the circuit, several iterations may be required until the simulated response matches the desired response. When this is achieved, work can begin on the physical design (layout) of the circuit. A layout editor is used to draw areas of n-type and p-type silicon, polysilicon and metal that will form the components and connections. The task is usually made easier by the foundry that provides parameterised cell macros that generate the layout data for MOSFETs, resistors and capacitors. However, for low-noise, well-matched, or compact circuits it is often necessary to create these devices by hand. The arrangement of devices and the connections between them is also carried out by hand. Once complete, design rule check (DRC), electrical rule check (ERC) and layout versus schematic (LVS) checks are performed to ensure that the final design is as intended. The design can be modified as required until a satisfactory conclusion is reached.

5.2. Digital System Design

A flow diagram of the steps involved in designing a digital circuit is shown in **Fig. 10.12**. In contrast to the analogue design flow, the physical design can be generated using software. The digital circuit may start off as a high-level "behavioural" model written in a programming language such as C or Matlab. It is then re-coded in a hardware description language (HDL) that allows the designer to describe digital circuits. The code is then compiled, simulated and debugged. Once the errors have been removed, simulation is required to ensure that the HDL code performs the functions described by the original high-level model.

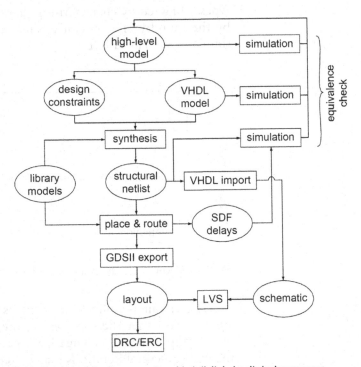

Fig. 10.12. Flow-chart for the computer-aided digital circuit design process.

If the HDL code uses only certain constructs that are allowed at a register transfer level (*see* (47) for a description of RTL), then the code may be synthesised. The synthesis process automatically generates the details of the gates and their interconnections to form a structural netlist. It attempts to optimise the netlist based on timing, area and power constraints that are set by the designer, and on the information contained in the library models provided by the foundry.

After synthesis, the netlist is converted into a physical layout by the automatic "place and route" tool. Delay information is extracted from this tool and used to annotate the structural netlist. If the simulation of this timing-accurate netlist fails to meet the specifications, then another iteration of the design loop is required. The next step in the process is to export the design, using the GDSII file format (the standard file format for transferring/archiving 2D graphical design data.) to describe the layout. The GDSII data and the HDL structural netlist are then imported into the layout editor (as used for the analogue design) to create the layout and schematic views respectively. As a final check DRC, ERC and LVS checks are performed as for the analogue design process.

6. The Wireless Environment

The transmission of radio signals in and around the human body is of considerable importance to the success of a wireless capsule device. The behaviour of an electromagnetic field in the presence of a human body is influenced by the dielectric properties of human tissue. In addition, the dielectric function is frequency-dependent and the absorption of electromagnetic waves increases with increasing frequency. As a consequence, reflection at boundaries, scattering, absorption and refraction of electromagnetic fields are frequency dependent. However, the radiation from electrically small sources in free space increases with frequency. In addition, the effect of the capacitive loading of the surrounding tissue on the source has a complicated frequency and spatial dependence.

Unsurprisingly, given the prevalence of mobile telephony, the majority of work that has been done to obtain a detailed understanding of radio transmission near the human body has focussed on the head and neck. Early work used relatively simple body models that assumed that human tissue was homogenous (48). As work progressed, research moved to more sophisticated models (49). More recently, there have been a number of studies looking at the abdominal region using detailed models for both female (50) and male bodies (51). The main results of these latter simulations are shown in **Fig. 10.13**. The simulations were carried out for an ingested transmitter at a number of possible locations and orientations and the data shown is therefore only representative.

Despite the fact that absorption of electromagnetic radiation increases with frequency it is found that, up to a point, increasing the frequency can improve the far-field signal strength from an

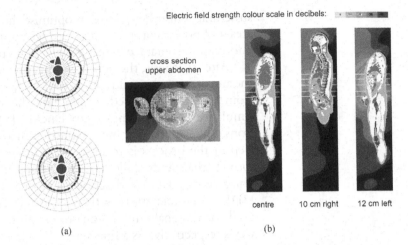

Fig. 10.13. (a) Far-field radiation patterns from an ingested source at 150 MHZ (*lower*) and 434 MHz (*upper*). The *solid line* is for E-field polarisation horizontal to the body and the *dashed line* is for the vertical polarisation. (b) the near-field pattern showing the field strength around the body.

ingested source (**Fig. 10.13a**). The reason for this is that the size of the capsule device demands that an electrically small antenna be used and that typically the antenna be very much smaller than $\lambda/2$ (the preferred size for the simplest radiating device), where λ is the wavelength. As a consequence, when the frequency is increased, the wavelength decreases towards the antenna dimensions, increasing the antenna's efficiency. It has been found that there is a competing effect between the increased efficiency of the antenna and the increasing absorption of the body tissue, and ultimately there is a trade-off in which an optimum frequency is found. Simulations have shown this to be in the region of 650 MHz which does not correspond exactly with any of the available ISM bands around the world. Since it is comfortably between European and USA frequencies, most major markets can be supplied.

The near-field pattern data is of interest since in many applications it is anticipated that the receiver antennae will be in close proximity to the body, which is for the most part located in the near-field for the wavelengths concerned. From **Fig. 10.13b** it can be seen that there is greater field strength to the left-hand anterior position of the abdomen due to the strongly absorbent nature of the liver on the right-hand side. The results of simulations of this kind can be used directly to assist in the design of the antenna system.

An alternative to conventional propagating wireless systems is to communicate wirelessly using the inductive near-field (52). Without dealing in detail with the electromagnetic problem here, it is possible to transmit wirelessly at low frequencies over distances that are very small compared to the propagating wavelength. In this evanescent regime, designers can take advantage of the lower absorption of RF power at lower frequencies and replace the antenna with two coils (one internally and one externally) that effectively behave as the primary and secondary coils of a transformer respectively. Detailed design and experimentation have shown that such a system can communicate effectively over the required range consuming less electrical power than a more conventional radio system.

6.1. Power Sources

Powering capsule devices is perhaps one of the greatest challenges. There are many possible micro-power sources currently being researched, a review of which has been written by Roundy (53). Not all the available techniques are useful in the context of an ingested device, e.g. solar power. More practical techniques include: a battery; electromagnetic induction; and electro-mechanical conversion. Other schemes, such as making direct use of gut mucosa as an electrolyte in an electrochemical cell arrangement, have been proposed but have not been explored in any serious way.

Although the use of batteries is by far the simplest power source, it comes with severe restrictions. Not all types of cell are favourable to use in implants (e.g. ZnO_2), and achieving adequate power density from the safer choices available is difficult (e.g. AgO). However a potentially greater problem than integrated energy storage is peak current delivery, since even short periods of high demand, for example during signal transmission, can very rapidly deplete a battery. As a consequence, when designing a microtelemetry system it is important to complete a detailed power budget and make design decisions that will ultimately compromise the devices performance in order to ensure correct functionality during a complete gut transit.

Electromagnetic induction is an attractive option not only because it reduces the power constraint on the device but also because removing the need for batteries would make the device smaller. The maximum power density permitted near the human body is in the region of 1 mW/cm^2 but varies from country to country (54). This is a severe limitation given the power requirement and distance over which power must be transmitted to reach a device deeply embedded in the human abdomen. With the exception of the relatively simple RFID temperature sensing devices developed for animal use that we have already discussed there has not yet been any significant demonstration of this technology for the more sophisticated human medical devices.

7. Packaging

Having decided upon the internal apparatus of a laboratory-in-a-pill it must all be packaged into a capsule. The package must be mechanically strong, chemically inert and allow access between the sensors and their environment. For optical devices there is always the prospect of an obstruction blocking the lens, but the clear plastic dome structures that are used in current products, such as the M2A from Given Imaging, are relatively easy to manufacture and strong. It is significantly more complicated to construct packages that will permit fluid access onto sensor devices, especially if these devices are integrated circuits or chips that will be adversely affected by current leakage due to liquids seeping into the encapsulating materials.

One of the main obstacles that has prevented the commercialisation of ISFET-based devices is the repeatability and reliability of the encapsulation procedure. It is normal for the encapsulant to be applied by hand, covering the chip and bond-wires but leaving a small opening above the sensing area. Epoxy is the most extensively used material although it is important to select a composition that is stable, is a good electrical insulator and does not flow during encapsulation. Many commercially available epoxies have been assessed for their suitability by making measurements of their electrical impedance over time (55–58).

By using UV-curable polymers, it is possible to increase the automation of the packaging process using a standard mask-aligner. A lift-off technique was developed using a sacrificial layer of photosensitive polyimide to protect the ISFET gates. Alumina-filled epoxy was applied by screen printing and partially cured, before the polyimide was etched away leaving a well in the epoxy (59). After ten days in solution, leakage currents as high as 200 nA were observed. Better results were achieved by direct photo-polymerisation of an epoxy-based encapsulant. ISFETs packaged using this method showed leakage currents of 35 nA after three months in solution. To avoid polarising the reference electrode, common to such devices, a leakage current of less than 1 nA is desirable (60).

This photolithographic patterning of the encapsulant was done at the wafer level to all the devices simultaneously. Subsequently the wafer was diced up and the individual chips were wire-bonded and coated with more encapsulant by hand. At the chip level, wire-bonded ISFET chips have been covered (again by hand) with a 0.5–1 mm thick photo-sensitive, epoxy-based film, then exposed and developed (61). After 20 days in solution, the devices retained low leakage currents. Some degree of automation was introduced by Sibbald (56) who used a dip-coating method to apply the polymers. They first recessed the chip into a PCB and wire-bonded the connections, before coating it with a layer of polyimide. Two layers of photoresist followed, before the underlying polyimide was etched away (**Fig. 10.14**). The slight undercutting of the polyimide was reported to be useful in anchoring the CHEMFET membrane in place. The packaged devices showed less than 10 pA leakage current after 10 days in solution. However, the encapsulation did exhibit electrical breakdown for applied bias

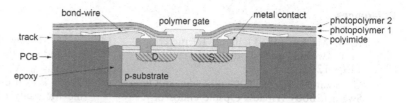

Fig. 10.14. Diagram of the cross-section through a CHEMFET device in a recessed PCB, encapsulated using a layer of polyimide and two layers of photoresist (56).

voltages in excess of 1.5–2 V. This was attributed to the high electric field in the thin layer of resist covering the bond-wires. More recently, a single layer of an epoxy-based photoresist (SU-8) has been used to package a pH-sensing microchip (62). In a separate study, photo-sensitive polyimide has also been used to create the wells that separate the ion-selective membranes on a multiple ISFET chip (63).

It is interesting to note that although flip-chip bonding is a well-established and robust packaging technique, it has not been applied to liquid-sensing ISFETs. In flip-chip bonding, solder bumps are patterned onto the bond-pads, allowing the chip to be directly connected to a PCB without the need for bond-wires. A gas sensor has been fabricated in this way, by bonding an ISFET to a ceramic substrate that had been coated with a suitable polymer (64). It may be that the high cost of solder bumping, which is normally applied to a whole wafer of devices, has prevented wider use of flip-chip bonding in capsule packaging.

8. Conclusion

Diagnostic pill devices have been in use for almost half a century and with the availability of miniaturisation techniques for electronics and sensors, they are now becoming practical devices for introduction into routine clinical practice. In this chapter we have outlined the history of the development of capsule devices. It is interesting to note that capsule developers have always been quick to adopt the latest discoveries from other fields, from the earliest germanium transistors to system-on-chip and laboratory-on-a-chip techniques. Of the current crop of devices, the video capsule in particular has drawn a great deal of attention, although pH-sensing devices for use in the oesophagus have perhaps made the greatest clinical impact. As new technologies develop and mature, we can only speculate as to what extent diagnostic capsule devices will become commonplace in hospitals, surgeries and perhaps even the home.

References

1. R. S. Mackay (1961) Radio telemetering from within the body. *Science* **134**, 1196–1202.

2. G. Iddan, G. Meron, A. Glukhovsky, and P. Swain (2000) Wireless capsule endoscopy. *Nature* **405**, 417.

3. G. X. Zhou (1989) Swallowable or implantable body temperature telemeter – Body temperature radio pill. *Proceedings of IEEE Fifteenth Annual Northeast Bioengineering Conference*, Boston, MA.

4. R. A. Casper (1992) Medical capsule device actuated by radio-frequency (RF) signal, USA patent #5,170, 801.

5. N. J. Clear, A. Milton, M. Humphrey, B. T. Henry, M. Wulff, D. J. Nichols, R. J. Anziano, and I. Wilding (2001) Evaluation of the Intelisite capsule to deliver theophyline and frusemide tablets to the small intestine and colon. *European Journal of Pharmaceutical Sciences* **13**, 375–384.

6. W. Weitschies, M. Karaus, D. Cordini, L. Trahms, J. Breitkreutz, and W. Semmler (2001) Magnetic marker monitoring of disintegrating capsules. *European Journal of Pharmaceutical Sciences* **13**, 411–416.

7. L. Phee, D. Accoto, A. Menciassi, C. Stefanini, M. C. Carrozza, and P. Dario (2002) Analysis and development of locomotion devices for the gastrointestinal tract. *IEEE Transactions on Biomedical Engineering* **49**(6), 613–616.

8. S. Guo, K. Sugimoto, T. Fukuda, and K. Oguro (1999) A new type of capsule medical micropump. *Proceedings of the 19th IEEE/ASME International Conference on Advanced Intelligent Mechatronics*, 55–60.

9. M. Sendoh, K. Ishiyama, and K. I. Arai (2003) Fabrication of magnetic actuator for use in a capsule endoscope. *IEEE Transactions on Magnetics* **39**(5), 3232–3234.

10. H. G. Noeller (1961) The use of a radio-transmitter capsule for the measurement of gastric pH. *German Medical Monthly* **6**, 3.

11. R. Lee (1995) *BSAVA Manual of Small Animal Diagnostic Imaging*, 2 ed., British Small Animal Veterinary Association.

12. P. B. Cotton and C. B. Williams (1980) *Practical Gastrointestinal Endoscopy*. Blackwell Scientific, Oxford.

13. L. M. Gladman and D. A. Gorard (2003) General practitioner and hospital specialist attitudes to functional gastrointestinal disorders. *Alimentary Pharmacology and Therapeutics* **17**(5), 651–654.

14. R. S. Mackay and B. Jacobson (1957) Endoradiosonde. *Nature* **179**, 1239–1240.

15. J. T. Farrar, V. K. Zworykin, and J. Baum (1957) Pressure-sensitive telemetering capsule for study of gastrointestinal motility. *Science* **126**, 975–976.

16. J. E. Pandolfino, J. E. Ritcher, T. Ours, R. N. Jason, M. Guardino, J. Chapman, and P. J. Kahrilas (2003) Ambulatory esophageal pH monitoring using a wireless system. *American Journal of Gastroenterology* **98**(4), 740–749.

17. Y. Sasaki, R. Hada, H. Nakajima, S. Fukada, and A. Munakata (1997) Improved localizing method of radiopill in measurement of entire gastrointestinal pH profiles: Colonic luminal pH innormal subjects and patients with Crohn's disease. *American Journal of Gastroenterology* **92**(1), 114–118.

18. A. G. Press, I. A. Hauptmann, B. Fuchs, M. Fuchs, K. Ewe, and G. Ramadori (1998) Gastrointestinal pH profiles in patients with inflammatory bowel disease. *Alimentary Pharmacology & Therapeutics* **12**, 673–678.

19. H. S. Wolff (1961) The radio pill. *New Scientist* **261**, 419–421.

20. B. W. Watson and A. W. Kay (1965) Radio-telemetering with special reference to the gastro-intestinal track. In *Biomechanics and Related Bio-Engineering Topics*, pp. 111–127. Pergamon Press, Oxford.

21. S. J. Meldrum, B. W. Watson, H. C. Riddle, R. L. Brown, and G. E. Sladen (1972) pH profile of gut as measured by radiotelemetry capsule. *British Medical Journal*, 104–106.

22. R. H. Colson, B. W. Watson, P. D. Fairclough, J. A. Walker-Smith, C. A. Campbell, D. Bellamy, and S. M. Hinsull (1981) An accurate, long term pH sensitive radio pill for ingestion and implantation. *Biotelemetry Patient Monitoring* **8**(4), 213–227.

23. D. F. Evans, G. Pye, A. G. Clark, T. J. Dyson, and J. D. Hardcastle (1988) Measurement of gastrointestinal pH profiles in normal ambulant human subjects. *Gut* **29**, 1035–1041.

24. L. Antoniazzi, H. T. Hua, and C. G. Streets (2002) Comparison of normal values obtained with the BRAVO, a catheter-free system, and conventional esophageal pH monitoring. *Digestive Disease Week 2002*, USA, abstract # M1700.

25. P. Swain (2003) Wireless capsule endoscopy. *Gut* **52**(Suppl IV), 48–50.

26. K. Kramer and L. B. Kinter (2003) Evaluation and applications of radiotelemetry in small laboratory animals. *Physiological Genomics* 13, 197–205.

27. G. Peters (1997) A new device for monitoring gastric pH in free-ranging animals. *American Journal of Physiology* 273(3), G748–G753.

28. J. M. D. Enemark, G. Peters, and R. J. Jørgensen (2003) Continuous monitoring of rumen pH – a case study with cattle. *Journal of Veterinary Medicine A* 40, 62–66.

29. ISO 11785 and ISO 3166.

30. R. A. Powers (1995) Batteries for low power electronics. *Proceedings of IEEE* 83(4), 687–693.

31. 'Recommendation 70-30 relating to the use of short range devices (SRD) (1997)' Conf. Eur. Postal Telecomm. Admin. (CEPT), Tromso, Norway, CEPT/ERC/TR70-03.

32. N. Aydin, T. Arslan, and D. R. S. Cumming (2002) Design and implementation of a spread spectrum based communication system for an ingestible capsule. *Proceedings of Annual International Conference IEEE EMBS.*

33. I. Nikolaidis, M. Barbeau, and E. Kranakis (Eds.) (2004) Proceedings of Ad-Hoc, Mobile, and Wireless Networks: *Third International Conference, Lecture Notes in Computer Science 3158,* Springer-Verlag, Heidelberg.

34. H. J. Park, I. Y. Park, J. W. Lee, B. S. Song, C. H. Won, and J. H. Cho (2003) Design of miniaturized telemetry module for bi-directional wireless endoscopes. *IEICE Transactions on Fundamentals of Electronics, Communications and Computer Sciences* E86-A(6), 1487–1491.

35. L. Wang, E. A. Johannessen, L. Cui, C. Ramsey, T. B. Tang, M. Ahmadian, A. Astaras, P. W. Dickman, J. M. Cooper, A. F. Murray, B. W. Flynn, S. P. Beaumont, and D. R. S. Cumming (2003) Networked Wireless Microsystem for Remote Gastrointestinal Monitoring. *Digest of Transducers* 03.

36. J. W. Gardner (1994) *Microsensors – Principles and Applications.* Wiley, Chichester.

37. A. Hierlemann and H. Baltes (2003) CMOS-based chemical microsensors. *Analyst* 128, 15–28.

38. C. Krüger, J.-G. Pfeffer, W. Mokwa, G. vom Bögel, R. Günther, T. Schmitz-Rode, and U. Schnakenberg (2002) Intravascular pressure monitoring system. *Proceedings of European Conference on Solid-State Transducers (EUROSENSORS),* M3C1.

39. H. Dudaicevs, M. Kandler, Y. Manoli, W. Mokwa, and E. Speigel (1994) Surface micromachined pressure sensors with integrated CMOS read-out electronics. *Sensors and Actuators A* 43(1–3), 157–163.

40. E. A. Johannessen, L. Wang, L. Cui, T. B. Tang, M. Ahmadian, A. Astaras, S. W. J. Reid, P. S. Yam, A. F. Murray, B. Flynn, S. P. Beaumont, D. R. S. Cumming, and J. M. Cooper (2004) Implementation of multichannel sensors for remote biomedical measurements in a microsystems format. *IEEE Transactions on Biomedical Engineering* 51(3), 525–535.

41. C. Hagleitner, D. Lange, A. Hierlemann, O. Brand, and H. Baltes (2002) CMOS single-chip gas detection system comprising capacitive, calorimetric and mass-sensitive microsensors. *IEEE Journal of Solid-State Circuits* 37, 1867–1878.

42. P. A. Hammond, D. R. S. Cumming, and D. Ali (2002) A single-chip pH sensor fabricated by a conventional CMOS process. *Proceedings of IEEE Sensors* 2002-1, 350–355.

43. B. D. DeBusschere and G. T. A. Kovacs (2001) Portable cell-based biosensor system using integrated CMOS cell-cartridges. *Biosensors and Bioelectronics* 16, 543–556.

44. M. L. Simpson, G. S. Sayler, B. M. Applegate, S. Ripp, D. E. Nivens, M. J. Paulus, and G. E. Jellison, Jr. (1998) Bioluminescent-bioreporter integrated circuits form novel whole-cell biosensors, *Trends Biotechnology* 16, 332–338.

45. Y. Zhang, K. K. Ma, and Q. Yao (1999) A software/hardware co-design methodology for embedded microprocessor core design. *IEEE Transactions on Consumer Electronics* 45(4), 1241–1246.

46. K. Kundert, H. Chang, D. Jefferies, G. Lamant, E. Malavasi, and F. Sendig (2000) Design of mixed-signal systems-on-a-chip. *IEEE Transactions on Computer Aided Design of Integrated Circuits and Systems* 19(12), 1561–1571.

47. S. Sjoholm and L. Lindh (1997) *VHDL for Designers.* Prentice Hall, Europe.

48. J. Toftgard, S. N. Hornsleth, and J. Andersen (1993) Effects on portable antennas of the presence of a person. *IEEE Transactions on Antennas Propagation* 41(6), 739–746.

49. M. Okoniewski and M. A. Stuchly (1996) A study of the handset antenna and human body interaction. *IEEE Transactions on Microwave Theory and Techniques* 44(10), Part 2, 1855–1864.

50. W. G. Scanlon and N. E. Evans (2000) Radio-wave propagation from a tissue-implanted source at 418 MHz and 916.5 MHz. *IEEE Transactions on Biomedical Engineering* **47**(4), 527–534.

51. L. C. Chirwa, P. A. Hammond, S. Roy, and D. R. S. Cumming (2003) Electromagnetic radiation from ingested sources in the human intestine between 150 MHz and 1.2 GHz. *IEEE Transactions on Biomedical Engineering* **50**(4), 484–492.

52. M. Ahmadian, B. W. Flynn, A. F. Murray, and D. R. S. Cumming (2003) Miniature transmitter for implantable micro systems. *IEEE EMBS Proceedings of Annual International Conference* **4**, 3028–3031,.

53. S. Roundy, D. Steingart, L. Frechette, P. Wright, and J. Rabaey (2004) Power sources for wireless sensor networks. *Lecture Notes in Computer Science* **2920**, 1–17.

54. Data presented is from the United Kingdom regulations for adults exposed to radiation in the band from 10 MHz to 60 MHz taken from http://www.who.org.

55. J. M. Chovelon, N. Jaffrezic-Renault, Y. Cros, J. J. Fombon, and D. Pedone (1991) Monitoring of ISFET encapsulation by impedance measurements. *Sensors and Actuators* **3**(1), 43–50.

56. A. Sibbald, P. D. Whalley, and A. K. Covington (1984) A miniature flow-through cell with a four-function CHEMFET integrated circuit for simultaneous measurements of potassium, hydrogen, calcium and sodium ions. *Analytica Chimica Acta*, **159**, 47–62.

57. A. Grisel, C. Francis, E. Verney, and G. Mondin (1989) Packaging technologies for integrated electrochemical sensors. *Sensors and Actuators B*, **17**, 285–295.

58. I. Gràcia, C. Cané, and E. Lora-Tamayo (1995) Electrical characterisation of the aging of sealing materials for ISFET chemical sensors. *Sensors and Actuators B* **24**(1–3), 206–210.

59. J. Münoz, A. Bratov, R. Mas, N. Abramova, C. Domínguez, and J. Bartrolí (1996) Planar compatible polymer technology for packaging of chemical microsensors. *Journal of Electrochemical Society* **143**(6), 2020–2025.

60. T. Matsuo and M. Esashi (1981) Methods of ISFET fabrication. *Sensors and Actuators* **1**, 77–96.

61. A. Bratov, J. Münoz, C. Dominguez, and J. Bartrolí (1995) Photocurable polymers applied as encapsulating materials for ISFET production. *Sensors and Actuators B* **24**, 823–825.

62. P. A. Hammond and D. R. S. Cumming (2004) Encapsulation of a liquid-sensing microchip using SU-8 photoresist. *Microelectronic Engineering* **73–74**, 893–897.

63. K. Tsukada, M. Sebata, Y. Miyahara, and H. Miyagi (1989) Long-life multiple-ISFETs with polymeric gates. *Sensors and Actuators* **18**, 329–336.

64. M. Fleischer, B. Ostrick, R. Pohle, E. Simon, H. Meixner, C. Bilger, and F. Daeche (2001) Low-power gas sensors based on work-function measurement in low-cost hybrid flipchip technology. *Sensors and Actuators B* **80**, 169–173.

SUBJECT INDEX

M.P. Hughes, K.F. Hoettges (eds.), *Microengineering in Biotechnology*, Methods in Molecular Biology 583,
DOI 10.1007/978-1-60327-106-6, © Humana Press, a part of Springer Science+Business Media, LLC 2010